"Looking for a fresh start? In *Dig. Lea,* unique path to self-discovery inspired by her bond with pets. ... not just a guide but a beacon of hope, offering practical exercises and heartfelt stories to enhance both your personal and professional life. Whether you're seeking fulfillment or growth, the wisdom gleaned from our furry companions will inspire you to embrace change. Get ready to unleash your potential—paws up for change!"

**Ernie Ward, DVM, CVFT**
America's Pet Advocate
DrErnieWard.com

"*Dig. Leap. Play.* offers a delightful journey into the parallel worlds of animals and humans, revealing profound analogies and lessons for living our best lives. Through captivating storytelling and insightful reflections, this book unveils the secret formula for discovering our true gifts and embracing a more meaningful existence. With its blend of wisdom and whimsy, it's a must-read for anyone seeking inspiration from the natural world."

**Cindy L Adams, MSW, PhD**
Professor, Veterinary Medicine
Director-Clinical Communication Program
Co-Chair Professional Skills
Adjunct Faculty-Cumming School of Medicine
& O'Brien Institute for Public Health
University of Calgary, Veterinary Medicine

"I was fortunate enough to be an early reader of Teresa's book, *Dig. Leap. Play.* I just happened to be in a stage of my life where its content touched me personally in a way that many of the self-development books I've read previously did not, and I've already started my "digging" work. As a result, I've also received the loudest "truth signal" I've had since I fell in love with writing.

Teresa is a highly skilled writer who knows how to get her points across in a non-judgmental way. She provides small steps, with the awareness that not everyone can simply turn their lives upside down to share their gift, and that's OK—a trait most self-development book authors don't seem to share."

**JaneA Kelley, Award-Winning Author,**
**Cat Blogger, and Mental Health Advocate**

"As a psychotherapist and executive coach, I find Dr. Teresa Woolard's *Dig. Leap. Play.* to align well with the principles I hold dear in my practice. This book thoughtfully applies the metaphor of our relationships with pets to the journey of personal growth, a perspective presented with refreshing simplicity, clarity, and playfulness.

*Dig. Leap. Play.* encourages the kind of introspection and step-by-step progress I advocate for in therapy and coaching. It offers readers a practical guide to understanding themselves better and making positive changes in their lives. What I appreciate most is its emphasis on actionable insights and the idea that small, playful steps can lead to significant personal transformations.

This book can be a valuable resource for those in my circle looking for a supportive nudge toward self-discovery and achieving their goals. It mirrors the belief in the power of personal effort and the importance of joy in the growth journey.

Please read it and enjoy!"

**Lindsay Tsang, Ph.D., RP, M.A. (Counselling)**
Owner of Lindsay Tsang & Associates | Psychotherapy | Counselling
Owner of Happy Hires | Strategic Planning | Executive Coaching
www.lindsaytsang.com

"*Dig. Leap. Play.* is a thought-provoking and inspiring self-help book with an interesting pet perspective on how to find our special talents, achieve our goals, and enjoy life. Dr. Woolard provides comprehensive information and structured steps on how to proceed and maintain motivation for success. It is written with kindness and enthusiasm that inspires confidence in achievement and is easy to read with personal examples to help illustrate. As a lifelong pet owner, avid reader, and self-help book enthusiast, I found *Dig. Leap. Play.* informative, enjoyable, and worth sharing."
**Amanda Burns, Mother and Grandmother**

"Teresa has the power of vulnerability and storytelling likened to that of Brené Brown!! Her ability to observe and connect with animal nature helps us better understand *our* lives and what we are truly capable of. Being able to make the emotional connection, is what I believe we all need more of."
**JenCB, M. Ed, Intermediate Teacher and Mindfulness Coach**

"This book was absolutely amazing! I found the book's exploration of finding one's gift through the wisdom and inspiration of pets to be truly captivating and thought-provoking."

**Taylor Harvie, Stay-at-home Mom**

"Dr. Teresa Woolard's book *Dig. Leap. Play.* is a beautiful meshing of self-discovery and the unconditional, non-judgmental way that dogs look at life. As I read the book, I fell in love with the connections. Anyone who is a pet owner will feel the same way. As I read her words, the dogs I have loved all jumped from the pages; different dogs for distinct reasons depending on the part of the book I was reading.

Being a Certified—Animal Assisted Interventionist Specialist and having had the opportunity to witness the beauty of dogs and students together and the power of the connection to transform lives this book resonated with me so much. I worked the exercises and found them to be fun yet revealing. I have found other self-discovery exercises to be daunting and overwhelming. These made so much sense and really allowed for an opportunity to dig into the layers of the onion of who I am. The playful aspect fulfilled. I believe this book and its companion workbook can be a fit for many audiences.

Dr. Teresa has lovingly created something that comes from deep in her heart. Her personal stories provide the reader with the image of where this work comes from and builds trust and rapport. Finishing the book for her mom, is so beautiful. For me it is a perfect book. I enjoyed a relationship with my mom that I miss every day. My absolute crazy love for animals, dogs especially, and my work in the canine-human bond world, made this book a safe space for me to work through my own stuff.

Thank you, Dr. Teresa, for your commitment to this book that I know will help countless others in their personal journeys to dig, leap, play, into their own joyful life."

**Lori Johnson, Teacher OCT**
Certified Animal Assisted Intervention Specialist
Author of *Set Up a Dog in School Program in 5 Days*

Dig.
Leap.
Play.

The Pet Lover's Guide
to Discovering Your **GIFT**,
Achieving Your **DREAMS**,
and Experiencing
**JOY**

# Dig.
# Leap.
# Play.

## DR. TERESA WOOLARD

Published by DeltaSpark Communications Inc.

ISBN (paperback): 978-1-0688513-0-8
ISBN (ebook): 978-1-0688513-1-5
ISBN: (audio): 978-1-0688513-2-2

Book design and production by www.AuthorSuccess.com

Image Credits:
Front cover image: istockphoto, Credit: Przemystaw Iciak
Maslow's hierarchy of needs: Shutterstock, Illustration Contributor: Apek Desi
Hippocampus image: Shutterstock, Vector Contributor: Blamb

This book is fondly dedicated to my mom, who was my best friend, cheerleader, and biggest advocate for this book, and who was also the most loving dog owner I have ever known.
Miss you, Mom xo

As a free gift to readers of my book, get a list of ten ways your pet can contribute to your well-being and development.
Click here to get the List.

10 Ways Your Pet Can Improve Your Health and Well-Being
https://drteresawoolard.com/free-list10

# Contents

# Foreword

**When Teresa asked me to write** the foreword to this book, I was hesitant. Yes, I'm a name in the animal community online given my former company, BlogPaws. And, I am a serious reader of self-help books, but I had to wonder, am I qualified to be a foreword writer?

My work as an editor and publisher seems to answer that question well, but a foreword writer is more than someone who just says nice things about a book. A true foreword writer is someone who cares about the book.

And there is a lot to care about in this book.

Truth is, I understand this book. It's full of the kind of insight and advice we need today. I say 'we' collectively because even if you think your life is all roses and sunshine, you must look inward once in a while and understand that some of those roses are wilting. And there are shadows where sunlight should be.

No one is immune to self-doubt or self-consciousness. We all have our troubles to bear; it's just that some of us recognize the need to learn how to improve ourselves, to find more sunshine and fresh roses, with the goal of finding our purpose in life and acting on it.

That's what you will find in this book. Your purpose. And you will learn how to act upon it. By understanding the power of pets and other animals to influence our emotional and physical health.

The writing is conversational. Inviting. Approachable. Teresa has written a book of advice that sounds like a few friends having afternoon tea. Over many weeks. Teresa purposely tells her own story to bring the reader into the dialogue—to help them feel comfortable making the changes she is suggesting.

Change is a prickly creature, isn't it? We don't like it much. Most of us. We want things to stay the way they are. Even if the way they are isn't giving us the result we want. We procrastinate, shrug our shoulders, and pretend everything is fine. "It's just fine," we tell ourselves. And then, when we look inward, which we must because our gut and our intuition are telling us to, and as Teresa wants us to, we discover things could improve. And while things might be 'fine'—what a trite word in response to the daily question, "Hi, how are you?" maybe 'fine' isn't good enough.

In *Dig. Leap. Play. A Pet Lover's Guide to Discovering Your Gift, Achieving Your Dreams, and Experiencing Joy,* Teresa has created a blueprint for happiness. Obviously, I am swayed by the references to dogs and other pets. I believe pets bring joy to our lives. I believe pets can teach us so much. I believe the studies that show pets help lower blood pressure and provide mental health support. I believe it because that's how I live my life.

But Teresa has gone further than that. She wants us to dig and dig deep to find those hidden bones that will help us change our lives. She wants us to recognize that animals can teach us how to calm down, get things done, turn ourselves around, and just enjoy every day.

The book is organized by the title: Dig first. Leap next. And Play last. These are the steps of the journey you will take as you learn more about yourself to find your true purpose.

In her introduction, she says, "Animals can serve as our guides on this journey in so many different ways. They have natural traits like stretching their limits, taking the leap, living in the moment, and enjoying the process."

As I write this, and as I read Teresa's words, my cat is sitting on my printer. She is giving herself an elaborate bath, stopping now and then to give me that look. The one that says, "It's lunchtime. Why are you still typing?" And this is what Teresa is talking about. How animals, those in our lives and those out in nature, live simply. They have their mealtimes (some inner clock we know nothing about), their playtime, and their time to sleep. They process their lives and their environment through those three portals. If all is well, they are happy.

As a veterinarian, Teresa has keen insight into animal behavior. Her studies over forty years have seen her witness changes in animal behavior that go back to the simple, ordinary way of living that nature has imbued in all of us but which we humans have forgotten.

The more I read *Dig. Leap. Play. A Pet Lover's Guide to Discovering Your Gift, Achieving Your Dreams, and Experiencing Joy,* the more I marvel at the power of pets to impact our lives the way they do.

Teresa tells us in Chapter 18, "Celebrate. . .Then Chill Like Your Cat:"

> The main reason I wrote this book was to get you to take action. If you don't take action steps toward your goals and what you want out of life, then chances are pretty good that you won't arrive there.
>
> Again, if I can show you a simple path to get there that's fun and manageable and reveal how others, including animals, could make it happen, perhaps you will take the necessary action steps to achieve your goals. In doing so, you could see for yourself that it's not as hard as you think, that you are capable, and you are special. Taking it one step further and having you focus on finding and using your gift, my hope here is that the action you take will be easier and more fulfilling for you because it involves using your gift. As I mentioned in the beginning, my belief is that we all have a gift, and our mission in life is to find, hone, and use it for the betterment of others.
>
> By reflecting on your journey, experiences, and learning and continuing to hone your gift, you'll be better able to serve others and feel even more joy and personal fulfillment.

It's my hope, along with Teresa's, that this book will help you understand and move forward to that ultimate goal of achieving your dreams and experiencing joy, every day. It opened my eyes to new ways to do that. It entertained me with stories and insights I am still thinking about today. Read it over time and learn to dig, leap, and play.

Yvonne DiVita
Master Book Builders
https://masterbookbuilders.com/

# Introduction

**Do you ever ask yourself, "Is this all there is?"**

Are you feeling stuck or unfulfilled? Do you feel like something is missing in your life, like more joy and passion? Do you want to make more of an impact but not sure how or what to do?

Then all I can say is, "Great!"

Great, because you are paying attention to what your body is trying to tell you. I believe your body, or your intuition, is sending you a wake-up call to change something about what you are doing, and instead do that thing that you are meant to do. Many people don't listen to what their intuition tells them and don't go after their "gift"–that thing they are good at and love to do (for many reasons). As a result, they miss out on the joy and fulfillment that using it could bring them. Ignoring this wake-up call could cause you frustration and even ill effects if you don't address this urging inside of you.

Welcome the frustration. Heed it and use it as motivation. If you picked up this book, I am going to assume that deep down, the frustration you are feeling is because you recognize that you are not using your gift, which is an important part of your personality and life. Use this frustration as motivation to fetch your gift; that thing you were meant to do and that will help you to live the life that you deserve. I'm going to suggest doing so by learning from the simplicity of animals.

I am going to show you how to find and use your gift once and for all . . . from pets! Yes, pets. Pets know what they want and how to get it, and we're going to model them and use their behaviors to help find your gift.

First, you are going to do a dig, just like a dog digging for a bone. I'm also going to share why it should be as important to you to find your gift as that bone is to a dog! Once you find your bone (I mean your gift), then you are going to work with it in some capacity. By doing so, the hope here is that you will enjoy using and sharing your gift just as much as a dog enjoys chewing on his bone. By taking action on your gift and stretching yourself in using it, you will experience more joy and fulfillment in your life and be as happy as a puppy or as content as a cat sunning herself on a window sill.

Animals just seem to have it right. They're more natural, instinctive beings that take life in stride. They know who they are, what they want, and go after it without second-guessing themselves. They live with ease and simplicity, and in the moment, they are ready for action and eager to play. Doesn't this sound like an exciting approach to living life?

Humans, on the other hand, are more easily led astray. We start out going to school and at a young age, have to decide what to do with our lives in our late teens to early twenties, often with little guidance and know-how. We choose a career based on some basic level of interest. We listen to our parents, teachers, and peers, and while they all are well-meaning, they often guide us based on what *they* know and want for us. So, we listen and heed them, thinking and believing that they know what's best for us. Then life gets busy. We tend not to have a chance to pause, reflect, and see if we're doing the thing that we are meant to do—"our gift"—until about our thirties and forties, or even later.

That is why I am so glad you picked up this book. This book is for you.

*Dig. Leap. Play.* will take you on a journey. A journey that will help you discover your gift at the stage of life you are in now. You will do a deep dive into who you truly are, find your special gift, and then take action. This is where the magic happens. It's only by using your gift that you will experience the joys it can bring by becoming more of who you were meant to be and feeling the exhilaration of finally waking up to your true self. By celebrating the action steps you've completed along the way and rejoicing in your progress, you will be motivated to not only keep going in reaching your major goal but to fetch future goals, too.

And don't think you are too old. You will also learn about a concept called neuroplasticity, in which the brain can continue to learn even in older age (thank goodness). So, the saying, "You *can* teach an old dog new tricks" holds true here both for our pets and for us!

Animals can serve as our guides on this journey in so many different ways. They have natural traits like stretching their limits, taking the leap, living in the moment, and enjoying the process. They are creatures of habit who can show us that relying on habits is much better than relying on our willpower. They also use laser focus, dogged determination, and a support team or "pack"—all of which are essential in finding and using your gift.

As someone who has studied and practiced veterinary medicine for over twenty-two years, with a special interest in animal behavior, as well as being a pet owner since age nine, I saw firsthand many unique traits and behaviors displayed by our pets. One of those traits that we see dogs using naturally is, of course, digging. The "dig" is the first step of the three-step process that you will learn and follow in *Dig. Leap. Play.* By doing a "dig" yourself, you'll be diving deep into discovering who you truly are at what stage in your life, and what special "gift" you own. This will set you on the right path to more joy and fulfillment because it will reveal a special part of you that perhaps has been buried for many years and whose discovery could set you free.

Other behaviors that our pets use that are simple, effortless, and natural will also be modeled in *Dig. Leap. Play.* like leaping and playing (to be discussed shortly), because why re-invent the wheel?

I also witnessed amazing changes in behaviors from bad to good in our animal friends using simple animal behavior techniques applied in practice, some of which you will learn in this book. It's not as hard as you think to make the necessary changes in your life and to develop the proper habits both physically and mentally that will be required to achieve your goal of using and finding your gift.

In addition to studying animal behavior, I have also "studied human behavior" through over forty years of being immersed in the personal development space. I observed common themes that were recommended for humans on how to make changes in our lives and achieve goals. Yet I found myself wondering how many readers, like me, fell into the trap of not taking the necessary action steps after reading a book and instead checked the box in our minds once the book was completed and placed it back on the bookshelf. There may be a lot of reasons we do this. Perhaps it's because the tasks were overwhelming, too complicated, not enjoyable, or because we didn't believe the results could happen for us.

*Dig. Leap. Play.* is different, because the action steps are simple, fun, and easy to do. They're not complicated nor overwhelming, and you will believe it's possible because, heck, if a dog can do it, so can you! A key focus in this book is to get you to take action. I believe that by modelling and applying the natural behaviors we see in our pets to the time-tested processes learned in personal development, we can finally not only find our gift but put it to use in a simple, fun, and natural way.

Learning from animals can serve as a reminder to live more simply, with a focus on doing more of those things that you are naturally good at and enjoy, like using your gift. You need to stop packing your days with meaningless activities that complicate and burden your life and distract you from doing what you should be doing. By using your gift, you can bring more fun and flow into your life and create a richer, more meaningful life experience.

The simple yet practical formula that you will learn in this book to help you find and act on your gift, will also serve as a tool moving forward that you can use for achieving other goals throughout your life. And I encourage you to do so. By going through this process once, learning the simple steps to take, celebrating your progress, and experiencing what you are truly capable of, you will continue to stretch yourself more and in other areas of your life. By continuing on this journey of using your gift and becoming more of your authentic self, you will live with more joy and passion, and may even amaze yourself in what you are truly capable of!

In this book, you will learn:

- How to ask yourself those hard questions and do a true assessment of who you are to ultimately uncover your gift

- Simple techniques to overcome hurdles, build determination and confidence, and be more focused

- How to create a plan to put your gift into action to share with the world

- The importance and value of play, and how to have fun again!

This book is divided into three main sections: Dig, Leap, and Play. After a brief introduction into why finding your gift is important, the first section is Dig, where you will learn to trust your intuition, uncover your gift, and create a Fetch mindset to get ready to take action in using it. The Leap section will have you taking action on using your gift by breaking down your "Stretch" goal into baby steps, creating daily habits, and using your helpers or "pack" to help you stay focused. The Play section reminds you to celebrate your wins, whether big or small, and to not only have fun, but rejoice in who you have become in this process. Each chapter will also highlight some amazing examples of humans and animals achieving the recommended stages along the process to engage, entertain, and inspire you.

There are exercises to do in the book called Fetch Steps at the end of each chapter as well as in the *Dig. Leap. Play. Companion Workbook* that is available online at companionworkbook.com. These exercises serve as reinforcements as to what is necessary to truly find your gift and put it to use. You will be stretched somewhat outside of your comfort zone, and it could get somewhat emotional for you, but I encourage you to trust the process, trust yourself, and get ready to "take the leap" to live a more joyous and fulfilled life by sharing your gift to the world.

I've used this process myself to experience many great things in my life, such as self-publishing children's books about pet care, launching products for the Canadian veterinary industry, training groups of hundreds of adults, opening two award-winning restaurants, and even competing in and winning awards in several fitness competitions in my fifties! I'll be the first to say that I'm nothing special, so if I can do it, you can too. I will go one step further and show you that if your pet can do it, so can you. I want you to live the rich, fulfilling life you deserve that will be memorable, impactful, and will live beyond you.

So, if you're ready to get started, then let's begin your journey to *Dig, Leap, and Play!*

# 1

## How a Gift is like a Bone

**I often ask myself, "What is our purpose here on Earth?"**

Other than procreating and keeping the human race going, why are we here? Surely it is more than working, raising a family, and gaining possessions–only to get rid of them later in life (not the family, but the possessions)!

I believe our purpose goes deeper than that. I believe that we all have a special gift and that our purpose here on Earth is to find that gift and use it or share it with others in some capacity to better humanity and the world.

What is meant by a gift in this sense?

A gift, described in the Google dictionary, is a natural ability or talent.[1] The Merriam-Webster dictionary describes a gift as a notable capacity, talent, or endowment; something voluntarily transferred by one person to another, without compensation.[2]

My own feeling is that a 'gift' is that special thing that you own or can do effortlessly, is in alignment with your values, and makes your heart sing when using it. You can do it tirelessly and you love to share it with others. It could be something that you possess, like a talent that you are born with, such as a beautiful, melodic voice with which to sing, or an "ear" for music to play guitar even if you have never taken lessons. I also believe it can be something that you acquire or hone over time based on the skills and knowledge that you gain over your lifetime, like writing, entrepreneurship, or leadership. Arnold Schwarzenegger exclaims in his documentary *Arnold* that his unique talent or gift is his ability to clearly envision in his mind what he wants. And he knows that when this vision is crystal clear, he goes after and gets that which he envisions.[3]

My family members have gifts of connecting with people and making them laugh, having an ear for music, being mechanically inclined, and performing in front of others. They are all different and unique for each, proving that even within a family, gifts can be diverse. Each family member uses their gifts in different ways, but still enjoys using and sharing them.

## In Your Element

When you are using or sharing your gift, you feel joyful and in flow. Opportunities start opening up for you and others comment that you are "in your element." This is what I see as your gift.

If you are not sure of what your gift is, don't worry. You will be going on a personal discovery journey shortly in the Dig section where you will dig deep inside of yourself, as well as get feedback from others who know you well, to uncover your 'gift.' Like a dog digging for a bone, we all need to be patient but also do the necessary work to uncover and reveal it. So, stay tuned and get ready. This chapter is all about showing you the importance of why you should bother investing the time to find your gift, and then once you find it, what you should do about it.

## Transfer Your Gift

To get the full effect from your gift, you should, as with any physical gift, transfer it from one person to another, whether for compensation or not.[4] This is where the magic occurs. Sharing our gifts makes us feel great. By sharing a piece of ourselves, through our gift, to better someone else or a group of people, we are leaving a legacy everywhere we go. This is the essence of truly living. This is something you should start now, regardless of your stage of life. Don't wait until you die to leave your legacy.

People can share their gifts in a variety of ways. Authors, for instance, share their gifts in books to affect their readers. Some write fiction and entertain. Others write non-fiction to educate. Speakers share their gifts through their voice and messages on stage or in a studio, and they do so to inspire, educate, and/or entertain. Artists can share their gift of painting or music through different mediums. Sharing one's gift doesn't have to be to the masses, either. It can be served one-to-one, like in a conversation or in a teaching session. Teachers or tutors whose gift is to teach and educate do so through courses, lectures, and seminars that can last days, hours, or minutes. The learning that takes place and the impact can be far-reaching. For some students, the impact could last a lifetime.

## Part-time versus Full-Time

Some people find that they enjoy sharing their gift part-time rather than making it their full-time occupation, and that is totally fine. The message here is to find your gift, use it, and share it in some capacity. No one is suggesting that you have to make major career moves here. Many of us can't or would rather do it one step at a time. By doing and expressing your gift, however, even if only on occasions or as a part-time hobby, the goal here is to feel the juice and fulfillment in sharing or giving to another. The recipient will value your gift and cherish the impact you've made on their life, possibly forever.

It also doesn't even matter how well you use your gift. It is not a world talent show. The act of simply giving a piece of ourselves to someone else is "doing our part" in society. How it is perceived by others should not matter. We cannot control how others perceive us, nor should we care. Focus instead on what you do have control over and that is your attitude, your actions, and your emotions. This book will focus on improving all three areas for you.

Some find their gift early while others find it later in life. Many unfortunately never find it.

For some, they are fortunate. They are born with a talent that they find and hone in childhood. They can put it to use and invest the time to be an expert by their twenties or thirties. You certainly have heard of people like this, and maybe even know someone personally who is like this. A good example here is Taylor Swift, who started singing at age thirteen and now at thirty-four has amassed a huge following, is at the top of her game, and has earned a billion dollars in the process.

Others may have a talent but were misdirected, usually not on purpose, but for some reason they went down a "more sensible" path. Sound familiar? Finally, some people will take time to develop their talent and will develop it through experiences and challenges in their life. The bottom line here is that it's never too late and anyone can find their gift at any time.

## Living a Lie

For those who never find and share their gift, they may feel at times that something is missing, but can't quite put their finger on it. (This was me in my late thirties.) Toward the end of their life they may ask, "Geez, I wonder if I lived my life fully? Was there something else I should have been doing with my life but didn't?"

Or worse, "Was my life a lie and I lived it for others rather than myself?"

Ouch. (I didn't *want* this to be me, which is why I feel strongly about this concept and want to help others who may be feeling this way, too. I don't want regrets at the end of my life and my hope here is that you don't, either.)

Australian author Bronnie Ware wrote in her book *The Top Five Regrets of the Dying* that one of the five common regrets that many of her patients told her about while she worked with them in palliative care was this: "I wish I'd had the courage to live a life true to myself, not the life others expected of me."[5]

But loving your life, rather than living a lie, is not going to happen for you just by reading this book. You will have to do more than just read this book and put it on the shelf. I ask and recommend that you do the action steps that I have laid out for you, because that is where the magic and confidence-building happen. The *doing* part is more important than merely reading the book and finding your gift. By keeping your gift to yourself, you risk missing out on the magic, the flow, the positive emotions, and the potential opportunities that can come your way and the impact you can make by using and sharing it with others.

## A Gift Gives Purpose and Worth

Finding and sharing your gift is important because it gives you purpose and worth. It allows you to feel the incredible power from the positive emotions that will bring you to life; emotions like joy, passion, love, inspiration, balance, and peace. You will feel and be motivated again. You will have a skip in your step and juice in your soul. Others will sense this about you and almost "feed" off of you. When you are being authentically you and giving of yourself through your gift, you are really giving from your heart. You feel complete when you give from the heart. You have done your "Earthly Job" in doing so, and are being rewarded as much as, if not more, than the person on the receiving end because you are living and leading the life you innately were meant to live.

## My Heaven

Can you imagine a world where everyone was living and doing their gift? People would likely be happier and more fulfilled. Everyone would be living in flow and enjoying what they were doing. There would be less stress and anxiety, and likely less depression. Employees would be looking forward to going to work and working productively, efficiently, and happily. Businesses would see better production, performance, and growth, and perhaps less worker shortage. More people may be working for themselves, using their gift as part of their business, aligned to their gift and values. The world would be a happier place overall, wouldn't you agree?

Industrial and Organizational Psychology Practitioner Adam Taylor shared in a LinkedIn post in 2019 about Parker Palmer's book, *Let Your Life Speak: Listening for the Voice of Vocation*, which exhibits a similar concept and the importance of finding jobs based on your gifts, values, and passions and the desire to leave a legacy or a mark on the world.

He shared that when the job doesn't 'fit' with these factors, employees tend to be disengaged from their work, resulting in lost productivity and morale, and will leave to find another one and continue to do this 'hopping' until they find the right 'fit.' This need to fit their gift to a job is especially prevalent among Millennials.[6]

## The Gift of Time

Following on this theme of career satisfaction but encompassing life satisfaction, too, a 2016 study about longevity was conducted by Allianz Life Insurance called, interestingly enough, the "Gift of Time." This study polled 3,000 Americans and asked them questions about their life to date and what they would do knowing that they may live thirty more years than their ancestors. One-third of those polled said they had regretted many of their major life decisions up to that point. Almost 40 percent wished they had followed through on their dreams or taken more risks with their careers. And 36 percent regretted not taking more risks with their lives in general, like trying something different that maybe would have stretched them from their comfort zone, taught them new skills and opened up new opportunities.[7] That's a lot of people who seem to have not found nor shared their gift, in my opinion. I will assume you don't want to be part of those statistics either, otherwise you wouldn't have picked up this book.

## Gift and Career

Could it be that people waste a lot of valuable time and chunks of their life drifting or staying in a job or career that doesn't fit simply because they didn't do the necessary work upfront to find their gift? I think so, and it speaks volumes as to the importance of taking the time and putting in the necessary effort to figure out *your* gift(s) and what you want to do exactly or before you make a career change. Taylor summarized this notion beautifully in his LinkedIn post mentioned above as this: "The clearer we are about who we are and what we want out of life, the better we can understand what work we should do and where we should do it."[8]

Doesn't this sound similar to what I mentioned in the introduction about animals–that they know who they are and what they want, and they go after it without hesitation? This is why we can use animals as our models. We can learn from them.

Why do so many not follow or bother to find their gift nor choose to use and share it? Is it because they don't have the time or don't know how to go about it? Do they not realize that they have one? Or do they realize they have a gift, but fear the change that may be necessary to carry it out? Are they worried that they would feel vulnerable and exposed? Are they lacking the confidence and know-how to take the leap of change? Do they fear that their financial situation will deteriorate by doing their gift?

## Falling Prey to our Perceptions

Who knows? Perhaps some or all of these have come into your mind or are even thoughts you have right now. You will notice that many of the above answers are mostly your perceptions and thoughts. And in terms of not having the time, I bet you could find the time if you made it a priority.

I believe that with productive time invested in this area, changing your perspective and building your confidence about yourself, and then taking the leap to using it, your life will hold so much more meaning and joy, and you won't have those regrets later in life. Is that not more attractive and meaningful to you than surfing meaningless photos and ads on the internet? Are you ready to make some changes that could light you up and make you feel more alive? Then read on.

Oh, and I'm going to say this here and now . . . I'm proud of you!

# 2

# The Training Process That's More Universal than You Think

**Why am I suggesting this notion** of learning from animals, or more specifically, our pets? Because I observed the simple animal behavior modification techniques that we use in veterinary medicine practice to create positive change and results in our pets. In simple terms, it worked! Did it work because it was simple? Yes. Did it work because the pet owner was determined to make it work and so he/she was consistent and focused on working with the pet? Certainly helps. Did it work because the veterinarian, and now the pet owner, better understood animal behavior? Most likely!

Once the pet owners had a better understanding about animal behavior, and how animals think and learn, it was easy to get their pets to do the things they wanted them to do, and not do the things they didn't want them to do. It takes patience, consistency, and time. But the beautiful thing here is that it's simple, fun, and easy, and this same philosophy can also work for people, and it has. What I want to highlight here for you is not to teach you Animal Behavior 101 (although you can learn that on your own), but to share some basic theories about animal behavior and why the simple techniques used with animals can work for you, too. Basically, I want you to be able to trust this process of learning from animals and incorporate some or all of these concepts into your own Leap Plan, your action plan which you will start in Chapter 10.

## Make it Habitual

Over my career and time as a pet owner, I have observed that animals like consistency. They are creatures of habit. For example, when training an animal, it's good practice to do the training in the same room and at the same time each day. They quickly begin to learn that this is the time that they get to work (learn) with and have time with you, and maybe even earn a treat! They start to anticipate this time together every day. Just like when they know when you'll be coming home from work, they instinctively know (through internal biologic rhythms) when you will be pulling into the driveway with your car. It's like they have a radar and can sense when you are near. The same holds true for "training time."

When I was training for my first fitness competition, I had to develop a habit in order to stay on track with my progress. The only time I could find to do my workouts around running my two businesses was from 9:00-11:00 p.m. daily at the gym. This became "my time." I didn't question it. It was simply what I had to do if I were to succeed at this crazy "midlife crisis" idea of mine of walking on stage for my first fitness competition. It became a daily habit for me (well almost; I did get one or two days off every week). I looked forward to this time and considered it my Zen hours (I actually missed it on my days off!). I'd put my earbuds in, crank some good tunes, and focus on my workouts. Weightlifting became my new passion, believe it or not. And this is from a girl who couldn't wait to get out of gym class in grade ten in high school! I started seeing the results, both externally and internally, in terms of energy, strength, and confidence. (By the way, I came in second place in my age category at the regional competition in Toronto, Ontario, and even competed with girls thirty years younger than me). So yes, becoming habitual can be a good thing because it works! None of this would have happened without creating those important good habits.

## Pavlov's Conditioning

I'm sure you're familiar with Pavlov's experiments with dogs from the early 1900s, in which dogs would salivate when presented with food, and then would salivate as soon as they saw the people that brought their food. Pavlov noticed this and changed his study from digestion in dogs to these "psychic secretions." When dogs were presented with the conditions that hinted that food was coming, they would salivate even when no food was presented. He paired a neutral stimulus, a bell, that normally would not create the response of salivation in dogs, with that of the presentation of food. After doing this many times, the dogs would salivate on the ringing of the bell alone.[9] This is called Classical Conditioning or Associative Learning, and it has been found to work across many different stimulus-response systems–for both animals and humans. By using this concept of stimulus and response conditioning, human behavior can be predicted and modified.

We can use conditioning to help us in situations when we may fear taking on a task that we may have been unconsciously conditioned to associate with negative feelings. Luckily, through consistent conditioning, we can change our perception about tackling that "fearful" task and make it a positive one.

Take the example of the gym. Let's say that you don't want to go to the gym because it evokes a negative feeling inside of you–perhaps pain or fear. This could be the result of something recent or from childhood. So, you avoid the gym. But by starting to take thirty-minute walks daily around your neighborhood, you will notice over time that this activity doesn't make you feel pain or fear, and that you are liking the benefits of weight loss, less stress, and feeling good about yourself. Once the mind accepts this form of conditioning due to the benefits, it would be ready to take more physically demanding tasks which would ultimately lead to viewing working out as a positive task rather than a negative one. And you just may find the gym less intimidating, even inviting.

The next time you find a task overwhelming, consider what stimuli perhaps may be making you feel that way about it. Once you have identified the stimuli, begin engaging in minor tasks that condition you to feel positive about the original task that you found daunting. Over time, your conditioning will lead you to change your mindset about your original task.

## Positive Reinforcement

Like dogs, we work for positive rewards. Both dogs and humans will work hard for a positive reward like, for example, delicious food. For most animals, where food is the reward, and they are hungry and food-motivated, we, as pet owners or trainers, can get their attention and they will be focused on doing whatever we want them to in order to get the treat. They will follow our lead, and with some guidance, do what is necessary to earn that treat. This was the same process I applied to myself when training for a fitness competition.

I had an online fitness coach who provided me with a meal and workout plan to follow for fifteen weeks leading up to the fitness competition date in Toronto. (I chose the last one for that year to give myself more time to prepare both mentally and physically.) At the end of each week, I would have to give her my progress notes, such as how I was feeling, describe any hurdles I had experienced, and take photos of my body. I worked hard during the week, focused on getting my reward, hoping that I had earned it. My reward wasn't necessarily seeing my body in photos (because quite frankly, I didn't notice the subtle differences week to week like my coach did). All I wanted was that four-ounce glass of red wine, a small steak, and that yummy piece of chocolate cake that I could have most Friday evenings if I had worked hard enough during the week to earn it. Like seriously, that was all I was focused on for the week! I wanted that small glass of red wine and cake so bad, my tongue tingled thinking about it. It tasted so good when it was earned and I felt I deserved it. (This concept of a treat or "cheat meal" here is to actually "trick" the body and increase its metabolism. Believe it or not, I would actually lose weight after my cheat meal!)

When it got close to the competition time however, and we had to "dial it down" so to speak in terms of my diet, these 'end-of-the-week' treats were sadly removed. This is called negative punishment, and it exists in the animal behavior world, too. When we remove something that the animal wants, the likelihood of that behavior occurring will be decreased. In training a dog to not jump up on a person, we remove the attention we give to the dog who is jumping up. He wants our attention, so by not giving our attention to the dog, he is not rewarded and will eventually stop jumping up. We have removed the thing he wants, which is our attention. Once he stops jumping and in turn sits, we then reward the desired activity, which in this case is sitting. (We had to do this exact method in our own family with our adopted dog, Waldo, who at first was an avid jumper. You will meet Waldo shortly.)

When I couldn't have my cake and wine, I felt deprived. I wanted to rebel actually, but I knew my coach was doing it for a good reason. I definitely did not want to look like the marshmallow woman (that would be punishment to me) on stage, so I stayed the course, followed the plan, and ate the bland food. But it is hard when you're working diligently at something and not being rewarded. You feel like you want to give up and throw in the towel. We'll remember this when planning your rewards in your Leap Plan.

In reality, my reward was (and should have been) to fit into my teensy-weeny bikini rather than devouring the chocolate cake, and truthfully it was. Many of us are wired, like our pets, to want an immediate reward, and when we don't get one, we may get derailed from our end goal and even want to give up. It's like a dog or cat stealing food from the table or counter. They know it's not desired behavior, but the immediacy of the reward is worth the risk, hoping the owner didn't see him or her steal it.

What works best in training both people and animals, in my opinion, is to focus on reward-based training or positive reinforcement; that is, rewarding the desired behavior. By rewarding ourselves along the path to achieving our goals, even in some small way, this helps to keep us on course and makes us feel good. We often make the mistake of not taking any reward until the final destination or goal is reached. The problem with that is that at times, like when I couldn't get my cake and wine, we want to stop and give up. We're naturally impatient. By rewarding ourselves along on the journey, even in some small way, it helps to keep us on track. I call these mini rewards, and we will be using these when you create your Leap Plan.

As a side note, positive reinforcement for people can also be in the form of gifts, incentives, recognition, and even compliments, not just food. Even our pets will work hard for verbal praise or a pat on the head.

Compliments go a long way to making people feel appreciated and, in many situations, will make a person repeat that rewarded activity or behavior, just like we see in our pets. For instance, if you were to compliment someone on their shirt or tie, chances are good that they will reach for that tie again soon, remembering the good feeling that he or she felt upon being complimented.

## Small Steps

Our pets, and I refer more to dogs (but in some cases, cats too), learn better when trained in small steps and with lots of mini rewards. We call this Shaping and Successive Approximation. We use successive approximation or incremental steps in teaching or "shaping" a new behavior for a pet to do. For instance, if we want to teach our dog to somersault, we can't expect it to learn how to do so in one step, mainly because it's not a natural move for a dog. We have to use small baby steps to help guide them to do what we want them to do, along the path to the end goal, which in this case is a somersault.

The pet owner or trainer would need a Shaping plan first, to know how and where to guide the dog, kind of like a roadmap of how to get it to do a somersault. Similarly, we humans need a plan in place so we know the baby steps along the way, to know that we are headed in the right direction and that the steps will eventually lead to the end goal.

In the case of the somersault, the first step could be to get it to touch its nose to a ball (the target). Using a food treat, guide the dog to the ball, and when it touches the ball, reward it with the treat. Then progress to lowering the ball to the ground, and reward when it follows the ball. Then bring the ball under its body, so it has to bend its neck to follow the ball. Next, have it bend its neck even further by bringing the ball under its back legs. Each time, reward the dog for doing the right moves that are desired. Eventually the dog will do a somersault.[10] Be patient, be quick, and keep it fun.

Again, we will use this Successive Approximation technique and mini rewards together in your Leap Plan to guide and keep you on track toward your goal, which could be a whole new behavior for you, just like a somersault is for a dog!

I hope I have convinced you enough here to trust the process that you are about to read in the chapters ahead about learning from our pets, mainly because if it works for animals, it should work for us! In creating your Leap Plan in Chapter 10, you will be taking advantage of many of these techniques that we just laid out in this chapter. Techniques like incorporating daily habits, using Shaping and Successive Approximation, and positive reinforcement. These techniques seem technical, but really, they're very simple and sensible. Making a change in your life, whether small or large, will be much easier using these techniques. I think you'll even find them fun.

So, I have given you some basic concepts here, but we need to distill it down to a simple formula. And that is what we will do in the next chapter.

# 3

# Dig. Leap. Play: The Secret Formula

**The formula that we will be using** in this book to help you find and use your gift is a simple three-step formula that I have coined: DIG. LEAP. PLAY. Why use a formula? The benefit of using a formula is that it's reproducible and easy to follow. Not only can you use it here in this book to find and use your gift, but to also go after any goal or dream that you have. It follows years of both animal and human research with a unique perspective of applying simple animal behavior techniques to the tried-and-true personal development space with an emphasis on keeping it simple and encouraging action.

Following is a description of each of the three sections of the formula:

## DIG

The DIG section will focus on you. It is about digging deep inside to pull out what you bring to the table now at this stage of your life–your attributes, knowledge, and experiences (i.e., tools)–that are collectively unique to you. You will be guided both in this book and in the *Dig. Leap. Play. Companion Workbook* (available at companionworkbook.com) to identify your strengths, weaknesses, values, passions, beliefs, hobbies, and interests. You will not only be digging on your own but also gaining feedback from others.

In addition to assessing and identifying what 'tools' you have personally in terms of your gift (or gifts), you will also be assessing your own supportive abilities as well as external resources you have available to you. Your own supportive abilities could be strengths or personality traits that you possess but are not strong enough to be your gift. However, they can help you in using and sharing your gift. For instance, having a determined mindset is super helpful when venturing out to promote yourself or a new business that uses your gift and not letting you stop in your mission to do so! External resources could be other people which we call your "Helpers" or your "Pack." These are people who could act as a role model, give you guidance on your journey, or simply cheer you on. We all need encouragement, and having Helpers on our side can truly make the difference for us in being successful, or not, in achieving our goals.

It's also important to identify and understand the positive beliefs and stories we tell ourselves about ourselves. These can make the difference between success or failure. As I mentioned earlier, it's often the thoughts we hold in our head that determine whether or not we leap to do something—whether it's to seek a promotion at our job, start a business, or write a book. Encouraging you to focus on saying and hearing more of the positive beliefs about yourself will be an inner helper for you and will serve you well to move forward, make progress, and to take the leap to achieving the goals in your life.

Likewise, being cognizant of the negative beliefs you hold about yourself is also helpful, because you will learn in this book how to deal with and replace them with positive ones. Using and leveraging these beliefs will give you the control, confidence, and power to succeed.

As part of the DIG process, you will also be identifying the hurdles that could get in your way of achieving the goal of using your gift. Unlike negative beliefs, which are personal and internal to you, meaning they come from within, hurdles are those things that are external to you, meaning they are outside of you. Hurdles can be responsibilities, people, or distractions. Recognizing the potential hurdles that can get in your way of going after your goal will allow you to better prepare in advance and have strategies in place to overcome them. You will be working on your mindset here to create a Fetch mindset. A Fetch mindset is one that will not allow the hurdles to stand in your way of getting what you want and living the life you deserve. A mindset not unlike that of a bulldog's mindset when digging for a bone, who doesn't give in until the bone is found!

Completing this DIG process could be very revealing for you. I hope that just by taking this first step, you will now have a much better idea of who you are and what you're capable of. By knowing your gift and using your Fetch mindset and belief system, you will be more confident and determined to go after those things that are aligned to your abilities and values. By doing so, you will start to enjoy life more, stretch yourself, and try new things. Who knows what opportunities could start to open up for you when you push your comfort zone, safely and in alignment with your gift and values? I'm going to suggest, as well, that when opportunities present themselves to you, and it feels right and excites you, then go for it. Because nothing happens without taking the leap.

## *"If you don't take the leap, you won't reap the reward."*
~ DR. TERESA WOOLARD

## LEAP

The next step in the formula is the LEAP. Here is where the rubber meets the road. This is where you take action on using your gift.

Once you've completed the DIG process and found your gift as well as your strengths, weaknesses, values, passions, beliefs, helpers, and hurdles, you'll be setting a goal on how to use your gift in some capacity. While yes, it's all fine and grand to find your gift, it's even better and more fulfilling to use it.

In the LEAP process, modeling some simple animal behavior techniques used in veterinary practice and a few life lessons from animals, you will begin on the path to using your gift and achieving the goals you set.

The first part of the Leap section will teach you how to create your own Leap Plan. This Leap Plan will serve as your guide in helping you to stay on track and focused on your goal. You will be taking baby steps–small, doable action steps–in the time frame that you set out. And where needed, you'll utilize your Helpers as resources and/or accountability partners. As much as you think you can do it all on your own, I'm urging you to use the power of accountability to your advantage. We will often accomplish more for others than for ourselves. When others are counting on or rooting for us, we will often stretch ourselves and work harder to please others over ourselves. This is also why so many successful businesspeople and athletes utilize a coach. We need guidance and encouragement. We need someone to help us stay on track. It's easier to give up on ourselves than it is to give up on the wishes that others have for us. The bottom line is we feel worse when we let others down compared to letting ourselves down.

Let me be your coach, if you will, in this book. Let your Leap Plan be your guide; your road map of what to do and when. However, it is still critical to use Helpers and/or accountability partners in this Leap process. Your chances of success will be much higher. The Dig Leap Play Facebook group can also serve as a Helper for you, with the other participants serving as reciprocal accountability partners, kind of like your Pack, and who understand because they are going through a similar process as you. In the DIG process, you will have assessed what Helpers you have available to you–whether people or resources. Use them to help you in your quest to achieving your goal.

What I noticed from the many personal development books I read, the courses I took, and the groups I joined–whether it was Toastmasters, business support groups, or network marketing–many participants failed to take action on occasion, including myself. I also observed that those that did take action often ended up being the ones that were successful. It was this noticeable lack of action, the missing puzzle piece for many, that was another key driver for me to write *Dig. Leap. Play.* It was also a complete contrast to what I saw in animals.

It was in my observation of animals and pets that made me ponder what it is that makes them just 'go for it.' What I concluded is that they are focused on what they want and don't overthink it. I felt that, within reason, perhaps we can take a page from our pets and learn how to do the same. Remove many of the distractions that we let happen to us (such as surfing the net and watching television) so we can focus better. Provide some simple tools, like the Leap Plan, to help keep us on track and not overthink things. By creating a Fetch mindset and setting out some simple tasks to follow, I believe, like our animal friends, we can accomplish more in our lives. By doing tasks or creating habits that use our gift and that are aligned with our values, we can enjoy the process once and for all and live with more joy and passion. This is my goal for you with this book.

The Leap Plan is the catapult for you in this book because it stresses the action piece. The key here is that the Leap Plan focuses on breaking down a large activity or task into smaller, more 'bite-size' steps (and rewards you for accomplishing them). This is what is missing for many of us in going after our goals and dreams and what will launch us into accomplishing what we set out to do.

If I can show you a simple and fun method to take action here, will you commit to doing it? If not, then the question to ask yourself is this: if not this method, or now, then when will you? Putting it off won't make it happen. I can't think of an easier and more fun method to help you to take action. Listen, if a pet can do it, do you see how it could be possible for you, too? Hopefully, you agree.

So, take the leap of finding, using, and sharing your gift. Help make this world better–one life at a time. Let's bring more joy, passion, and love to this world, your community, your family, and your friends. Let's be more like puppies and bring more smiles to the people we know and who get to do business with us. Let's give more of ourselves both naturally and wholesomely and make the world a better place in doing so. Leave your legacy by making an impact through shining and sharing your gift. Let others enjoy it, and at the same time, feel great yourself. The rewards will be exhilarating, exceptional, and worth every second that you invest in working your Leap Plan.

Are you ready to commit now?

## PLAY

Speaking of rewards, the third and final step of the formula is PLAY.

Why PLAY? Simply because we need more play in our lives and play is rewarding. We work hard. Some of us work more than one job to make ends meet. Costs have escalated in a variety of areas of our lives. Even after getting through the pandemic, the challenges of just living, managing, and surviving continue for many. We get focused on appeasing and doing for others, like taking our kids to their extracurricular activities or doing the seemingly endless maintenance-type tasks around the house, that there's little time left in our week for us to enjoy ourselves, or to play. We also don't take the time to celebrate our wins, whether big or small. That needs to change. I'm going to highlight the many benefits of play here, which is why it's a part of the process.

In the animal world, we use rewards or treats *a lot* to get our pets to do something that we want them to do or simply to create good habits and behaviors in them. It's amazing what a pet or a person will do for a reward, treat, or praise. Again, why not emulate and do the same for humans? No need to reinvent the wheel when we have a simple model to follow.

Utilizing a reward system, or what we refer to in veterinary medicine as positive reinforcement, goes a long way even for us. Positive reinforcement for humans helps build confidence, injects some much-needed enjoyment and excitement into our lives, and simply makes life more fun.

The essence of a reward is to give us something to strive for and to have a little win in our lives. Getting a reward creates a burst of dopamine, one of the "happy hormones" in our brain that internally makes us feel good.[11] Dopamine is responsible for allowing you to feel pleasure, satisfaction, and motivation. It changes your brain's biochemistry and emotional makeup to one that supports you positively. So why not take advantage of this, right?

Animals will do just about anything for a treat, especially if it's a tasty treat and if they are food-motivated. In practice, I used to wear a little fanny-pack-type pouch around my waist that was filled with dried liver treats. I would give a treat to a dog upon entering the front waiting room of the veterinary clinic for the sole purpose of having the dog associate the veterinary clinic with a positive treat. If every time the pet came in and received a treat, chances would be good that he or she would equate the clinic with a positive experience and thus not fear the veterinary clinic. It worked! Many pet owners would recount how the pet became excited to come into the clinic, whereas before it was reluctant. Treats can do magic!

We used treats in the examination room too. We would often put a few tasty treats on the exam room table for the pet to eat while being palpated or peanut butter on a finger for it to lick while the veterinary technician took its temperature or cut its nails. Some animals were too fearful to eat, but for many, this worked wonders. The experience of being in the exam room through using a treat system allowed for the experience for the pet, and pet owner in some cases, to be somewhat less scary and something not to be feared.

Similarly with humans, if we are working on a goal and getting some mini rewards (or treats) along the way, it can encourage us to continue. We will enjoy the process more and stepping outside of our comfort zone will not be as scary. It's important to include in your LEAP Plan those rewards that you like and that will drive you, otherwise you may not work as hard to get them.

While you will be instructed in creating your Leap Plan to set a big reward for yourself in achieving your goal, there will be smaller or mini rewards that you will include as well in your Leap Plan as you go along on your journey to achieving your goal. Again, these little rewards serve to keep you going and to keep it fun. Please don't ignore this step.

Play doesn't have to refer to just material goods, either. It could mean a celebration or a lunch meeting with others. Have your Helpers or accountability partners, if they are local to you, join you in celebrating a milestone that you've achieved. This will be fun for both or all of you!

While incorporating both mini and major rewards in your Leap Plan, it's important to keep focused and enjoy the process of taking action. While you will enjoy receiving the major and mini rewards, and so you should, it is important to note here that the biggest reward for you in partaking in this Leap Plan will be the changes in **you** by going through this journey. You will be stepping more into the 'real you.' You will become more confident about who you are and be more self-assured. People may even start noticing the changes in you and comment along the way. Take in and honor any compliments you may receive while on this journey. These are also rewards, albeit unexpected ones (and sometimes the best kind), and you deserve these 'treats' too!

They may notice that you're smiling and laughing more, or that you seem happier. These are all signs that you are on the right path. In my opinion, these should be considered BIG WINS because isn't this ultimately the real goal for us all? To be happier, more fulfilled, and living more to our true selves?

Rewards could be in the form of a new outfit, a massage, or a trip. It could be a night out for dinner, a movie with a friend or your spouse, or a pedicure. You get to decide what rewards will push you to complete the action steps you've laid out in your Leap Plan.

Do not downplay the importance of rewards and play in your Leap Plan. Rewards are appealing to the animal brain of ours that seeks and benefits from the positive endorphins that come from receiving rewards and positive compliments. Do not shortchange yourself here. You do deserve them, especially if you are taking the necessary action steps. Rejoice even from the small wins. Your body and mind will love it! Do whatever you can to nurture and satisfy those basic human needs of loving yourself and feeling good.

Just like the animals in the clinic, the positive rewards will make this entire process a more positive experience for you. They will keep you doing whatever is necessary to achieving your goal and using your gift. Then, eventually you will be able to experience the big reward for you, which ultimately will be achieving your goal of using your gift, becoming the person you were meant to be, and living a fulfilling life because of it. The process is that simple. To summarize, there are three easy steps:

- Dig–to better understand yourself and find your gift as well as your Helpers and hurdles;

- Leap–to take action to go for it and find ways to use your gift; and

- Play–to reward yourself along the way and enjoy the journey in finding, using, and sharing your gift.

However, before we start into the process of doing your Dig, I want you to learn about trusting your intuition, or what I like to refer to as your animal instincts. Some would argue that they are different, with animal instincts being more basic, but the animal analogy here is what we are using and leveraging, so it works in this context. Plus, we're trying to keep things simple and basic anyway.

I'm excited for you to read the next chapter about Trusting Your Animal Instinct. This is what triggered the turning point for me, which is why I believe so much in its power.

## FETCH STEPS

1. Start thinking about some rewards that would be enticing enough for you to want to work hard to get.

2. Categorize the rewards you listed in Fetch Step #1 into major and mini-rewards. You will add some or all of these into your Leap Plan in Chapter 10.

# DIG

# 4

# Trust Your 'Animal' Instincts

**Going further on this theme of knowing** ourselves and tapping into our inner world, we should be turning to our bodies and trusting our feelings and emotions, in addition to our minds. Here's why.

We differ from animals with our minds, our thought processes, and thinking abilities mainly because we have a more developed neocortex. While this major difference is beneficial for us in many ways, like being able to speak, perceive, remember past personal experiences, and to think ahead and in abstract terms, I wonder if we are thinking too much? Are we over-thinking, like, for instance, when we want to take action on something that we aren't familiar with? Are we using and depending too much on our rational minds alone without taking into account our intuition, our instincts, our bodies, and how we're feeling?

Our bodies know a lot about us, even more than our minds, yet we seem to have lost the art of listening to our bodies. We rely mostly or solely on our minds, which isn't always the best solution because our minds can be filled with negative and limiting beliefs. You will discover these in the DIG process in the next chapter. Many of these negative beliefs we have heard from a young age from our parents, teachers, peers, and the media. These beliefs have the capacity to effectively and permanently discourage us from ever looking inside ourselves for guidance, searching for what our hearts want, or what our true abilities and strengths are. Even if we did search inside ourselves, would we trust ourselves to move forward? Are we confident enough in our own intuition? I think not, in most cases.

## Know Yourself

We have all grown up in a society that is filled with external influences ranging from family, friends, society, culture, religion, the media, and now the internet. We are barraged with messages that come from these sources, and often times these messages can have an impact on how we think and feel about ourselves. We often rely on these outside sources of information for advice or solutions to our problems. We look to them to tell us what to do and what not to do. We follow their well-meaning advice, playing by society's rules and doing what is expected of us. The problem is, we are not necessarily living our true selves or with our true selves' needs, desires, and gifts in mind. We aren't doing so because we don't *know* our true selves. We forget to trust ourselves and our own intuition, and instead put our trust in the opinions of others. This can lead us down a path of self-doubt, confusion, and anxiety.

I believe this is because we simply haven't taken the time to get to know our true selves and trust our own feelings and intuition about ourselves. It is easy to lose ourselves and fall into the trap of living our lives for others and by others' wishes, and not realize it until much later in life (or worse, not at all), when many may say at that point that it's too late to change.[12] I say, "Nonsense." It's never too late to find yourself. Once you realize that you may be in this situation, it is imperative for you to take the time to trust your intuition. Just like Socrates said years ago, the secret to happiness in life is to "*know thyself*."[13] Start by listening to your body. You will be much happier that you did.

To learn how to listen to our bodies, let's look at our animal friends for guidance. Because essentially that is what this book is all about–learning from our animal friends. Animals have not forgotten how to listen to their bodies. They instinctively know how to live and what they need. It is their instincts that drive them. They use them to stay safe and to catch prey. Here is a quote that sums it up: "*A dog is at its happiest doing what its instincts tell him to do*."[14]

Here again, we can learn from them. Perhaps if we listened more to our 'instincts,' or our intuition, we would be happier, too!

## Intuition

Whether we call it intuition, an inner voice, a psychic power, or vibes, animals and humans share this capacity to respond to their environment in ways we know little about. Animals appear to have a more finely-tuned sixth sense compared to humans. The good news is that we can learn to tap into our intuition, this sixth sense of ours. We can learn to listen better and be more aware of our bodies.

How are you reacting in certain situations? What emotions are you feeling at any given time, whether at work, home, or play? Are you feeling stressed? Are you experiencing a tightness in your neck or gut or are you warm and elated all over, as if ready to jump out of your skin? These are signals that your body is sending you to give you a message. We need to embrace and trust these bodily signals. Treat your body as an ally, rather than something to control or push away. Let your inner knowing serve as your guide. (More on that shortly.)

It has been said that the "intuitive state is the natural state of your soul."[15] How we feel about experiences in life dictate how we feel about life itself. So, to feel happier, we need to experience more happy moments. We need to do more of those things that give us those happy feelings. Conversely, we need to do less of the things that create negative or anxious feelings so we can feel happier.

If you do something and it doesn't feel right or you get a pit in your stomach beforehand or during the task, it could be your body telling you that whatever you are doing is not right for you. Perhaps you should stop what you are doing, or at least reevaluate the situation. It just may be that it's not safe to continue, or perhaps what you are doing is not congruent with your values, or it's simply too uncomfortable for you.

I pride myself on being in tune with my body. In fact, it was a body prompting that finally catapulted a change in career for me and a major move for our family.

## The Rat Race

One morning years ago, as I was driving to work, taking the same route I had taken for months, I suddenly asked myself, "What am I doing this for? Why am I riding this treadmill? I am not having fun."

I had just taken my two young children, aged three months and twenty-one months, to the daycare, crying afterwards, which I did daily, feeling guilty that I was choosing a career over them, which wasn't the case. I simply felt compelled to do "my share" to contribute to the family income, helping pay for our mounting expenses of a mortgage, two cars, and two children in full-time daycare. It was only logical and expected that I should be doing what I went to school and trained for, right?

I was feeling the trap of doing what society expected of me. Doing the "shoulds" rather than the "wants." Sound familiar?

That morning, while driving to work, it hit me like a ton of bricks. Why was I feeling more like a rat on a wheel, trying to keep up with the Joneses, than the professional I was supposed to be? I was feeling anxious about heading into work, as I had every morning, stressed about what I might be facing that day. I worked at a very busy clinic requiring a frenetic pace just to keep up. I was lucky enough to grab twenty minutes for lunch during the day. Practicing veterinary medicine was definitely different from what I had originally imagined it would be like. I dreaded going into work. The door of the clinic felt like a 200-pound steel door that I hated opening and walking through. There were times during the summer months when I would volunteer to pull weeds in the back gardens by the parking lot just to get a break from the mental and physical stresses of the clinic environment.

After several months of enduring repeated bouts of upper respiratory illnesses, being sick often, losing twelve pounds in a year without even trying, and being bedridden with suspected pneumonia, I knew I had to make a change. I feared something worse could happen to my health if I didn't make a change. This was a signal from my body that enough was enough. I could not continue this pace, conflict, stress, and uneasiness any longer for fear my health would suffer even more. Being a normally healthy person who was rarely sick, this realization was big news for me. I'm not sure why I didn't see this earlier. Perhaps because I was just going through the motions, doing instead of listening. I guess I just finally 'woke up.'

To many on the outside, it looked like we "had it all." A large home in a very desirable area, a happy marriage of professionals with good-paying jobs, with two children in daycare. The truth was, I was miserable. I was simply not happy with what I was doing. I learned a lot about myself from this process. I learned that for me, money cannot buy happiness; being congruent with your values and loving what you do makes one happier than acquiring all of the material trappings that we are told we should want. Steve Jobs said it best at the commencement speech he gave in 2005 to the graduates of Stanford University. He advised, "Find what you love . . . love what you do."[16]

I learned that I needed more than a paycheck to feel fulfilled and happy. When I started my home-based veterinary marketing business later, I made less money initially, but for me, that didn't matter. My health and family were far more important to me than making a certain income. I also discovered that doing what I innately want or need to do pays much greater dividends in terms of my emotional health, life satisfaction, and fulfillment than what a paycheck alone could ever offer.

## Make a Change

I knew that things were not quite right for me. They weren't congruent with my values or innate abilities. I realized that I needed to do a self-assessment (a.k.a. a DIG) to make some changes in what I was doing; to delve deeper into what I wanted and needed and find something that would be more in line with my true self. I realized that what was missing in my life was my ability to create.

Later that night, I discussed this realization with my husband. Being the supportive, caring individual he was, he agreed that we needed to make some changes. We decided to sell our home and move to a less expensive place that would allow me the freedom to practice part-time so I could focus on my home-based veterinary marketing business which I had dreamed about starting for a few years at that point. My husband had been in pharmaceutical marketing, and I used to love the creative, thought-provoking, and professional marketing pieces his company would develop for the human medical field. I thought I could bring not only a veterinarian's perspective to the promotional pieces used by pet food and animal health companies, but also my creative skills to the Canadian veterinary industry.

Even though we moved to a smaller town with no family nearby, it was the best decision we ever made, and we have never regretted the move, even twenty-three years later. I rarely got sick, was much happier with my life, was more available for my kids growing up, and loved the flexibility that my new situation offered.

By finally listening to my body and making some changes, albeit major ones, I now had more opportunities to be creative, was less stressed, and felt much healthier and happier. From this experience, I learned the importance of being in tune with my body and to take notice, earlier rather than later, of any signals, good or bad, that my body sends me. I also learned the importance of taking action on these feelings rather than just accepting them and letting them slowly deteriorate me. Taking action by making changes to improve my situation brought improvements in my emotional state, perception of my situation, and to my overall life. I hope to impart some of these learned habits to you here in this book.

## Getting in Tune

So how do you get "in tune" with your body? As a definition, first of all, getting 'in tune' with your body is really about being aware of, and accessing, your intuition. Intuition is an important resource that can greatly contribute to your success and fulfillment in life. It is a practical tool to help point you in the right direction at any given moment and successfully deal with life. It is like a filter or guide by which you can assess if what you are doing is the right thing for you or not.

The good news is that we can learn to get in touch more with our bodies and our intuition. Intuition is a natural thing that we are all born with. It's just that our Western culture emphasizes the development of the rational mind, paying little attention to the value and existence of intuition. As a result, many of us have lost touch with our inner selves.

There are some great resources about developing your intuition that I would encourage you to read. One is *Developing Intuition* by Shakti Gawain. Another is *The Intuitive Way* by Penney Peirce.

## Develop Your Intuition

Developing your intuition, as Shakti Gawain believes, is one of the most valuable things you will ever do for yourself.[17] Penney Peirce in *The Intuitive Way* says, "Intuition can bring increased success and satisfaction in every realm–be it material, emotional, or spiritual–and can also bring many joys, both tiny and great."

Overall, intuition can make your life smoother and more fun.[18] And isn't that what we all want? To make our lives simpler, smoother, easier, and more fun? I know I want that (especially in the mid-life years in which I find myself these days).

Some would describe this type of life as 'going with the flow.' By listening to, accessing, and trusting your intuition, you start living in the flow. Opportunities start opening up for you and life becomes easier and smoother, requiring less effort, it seems.

While part of tapping into your intuition is like a form of meditation in a relaxed state, other methods involve learning to be aware of our bodies' signals. While not much has been written about so-called body signals, most of them are common sense. Here I have listed some common positive and negative body signals that I experience personally or notice in others who are experiencing them.

## Positive or Truth Signals

Your positive or truth signals are revealed when you are in synch with your intuition, while your negative or anxiety signals are a reflection of not being in synch with your intuition. Have you experienced any of these? Are you feeling them consistently at work? At home? Or on certain days of the week, like Monday mornings?

## POSITIVE OR TRUTH SIGNALS

a general feeling of happiness

energetic

warm feeling all over your body

content

relaxed demeanor

laughing

smiling

light mood

enjoying life

optimistic

youthful look and feeling

pep in your step

overall calmness

enthusiastic

tingling up the back

heart wanting to expand out of your chest

# Negative or Anxiety Signals

Negative or anxiety signals are those signals that make us feel bad, anxious, or unhappy. I often felt sick to my stomach or couldn't eat when I was anxious about going into a stressful situation where I didn't feel on top of my game. My gut would continually be "turning over," and I often got diarrhea when I was stressed or anxious.

## NEGATIVE OR ANXIETY SIGNALS

depressed

sad

negative mood

feeling frustrated

unhappy

head low

dissatisfied

stomach in "knots"

"pit" in your stomach

increased gut motility

diarrhea

feeling ill

lethargic

tired-looking, aged look

tightness in gut or back

flushed face

short-tempered, "snappy" mood

venting on others

These signals are messages from your body providing feedback to you. Your truth and anxiety signals are your pipeline to your higher self. Your body is constantly communicating with you about every option or action you consider. Most of us simply do not take the time to notice, yet our bodies are speaking volumes. We're just not paying attention.

We often miss the body's nonverbal cues of communicating because our minds are so preoccupied. Our minds are constantly being bombarded with ongoing distractions such as ringing or vibrating phones, text messages, emails, social media, work, and home demands, that there is little time remaining for our minds to be quiet so we can focus on our bodies. Animals don't have the materialism and stresses of today's society to fill their heads with, so they can focus more on their animal instincts. This point emphasizes the importance of quiet time as well as decreased reliance on material goods. We need to make a conscious effort to set some quiet time aside, ideally on a daily basis, to listen more to our bodies and be more in tune with our feelings. Make a date with yourself. Interpret your feelings, stop and ask yourself, "How am I feeling right now? How was I feeling today?"

## Think Like an Animal

If we are looking for more meaning and personal fulfillment in our lives, then we need to pay attention to what our bodies are trying to tell us about certain situations we may find ourselves in, about people we meet or are with, or even about certain jobs or careers we are doing. Being comfortable with your feelings is important in connecting to your inner guidance; your direct knowing. Emotions are information.

If you want to improve your intuition, simply move down the chain of knowing–from your head to deep inside your body, into your solar plexus, a network of nerves behind the stomach that sends and receives messages to and from local organs as well as to the brain.[19] [20] Go below words from your logical brain or rational mind and delve into your senses. Think more like an animal. Once we learn to listen to our body's cues and learn to decipher the messages sooner, we can take advantage of opportunities that create happiness and not waste unnecessary time and energy on things that don't fit with our values.

Intuition is like a prompting from your body or a gut reaction. Don't ignore it. You can forget to pay attention to it, but you can never lose it entirely. It will continue to nag at you until you finally pay attention to it. I certainly learned this fact. For me, my life improved immensely once I took the time to listen and made some changes for the better that were more in line with my values and needs.

## Trust Your Intuition

Once you tap into, trust, and follow your intuition, it will create feelings of elation for you. You will feel at your best. Others will notice and compliment you. You may start getting feedback from others like "You look great," or, "You have a glow about you!" or, "You make things look so easy."

Such feedback is an indication that you are living in synch with your intuition.

Let your intuition be your guide when making decisions. It knows you and it won't steer you wrong. Intuition can show you step by step what you need to do to fulfill your heart's desires and achieve your goals. Following your intuition will put you 'in the flow,' which is a very productive and desirable state. When in the flow, things go well and smoothly, and we feel energized. On the other hand, when we do not follow our intuition and we are not in the flow, things seem more difficult to achieve. We feel frustrated, depleted, or even depressed and numb. Life can seem like a constant upward hill that we're struggling to climb.

When you are in sync with yourself and receiving intuitive insights regularly, your creativity flows, and life seems effortless. You may even remark, like I have when this happens, that "Boy, everything just seemed to fall into place." This could be a sign that you are experiencing "flow." The more you ask your inner guidance and act on your intuition, the more your outer life will reflect your inner thoughts, beliefs, and commands. Then you will be living your true, authentic self. Your confidence will soar, and you'll feel triumphant!

## Tap Into Your Intuition

To tap into your intuition, your inner knowing, first make some quiet mind time for yourself. Ideally, find a quiet place with little to no distractions. Get comfortable, either sitting upright, cross-legged, or lying down. Some intuitive counselors suggest having both feet on the ground to receive the earth's energy. Just like with meditation, close your eyes and focus on your breathing. Don't let your mind wander onto your tasks for the day. Simply focus on your breaths. Imagine your breath flowing in through your nostrils and out. I imagine this gentle, swirling, golden colored wind coming in my left nostril, flowing into my lungs and down into my stomach area. Then I imagine this swirling gentle breeze exiting my right nostril. I refer to this wind as my 'golden guidance.' This visual imagery helps me stay focused.

After four or five slow breaths (provided you haven't let your mind wander onto tasks you need to do for that day), then focus your mind inside your gut or solar plexus area. Ask (silently or out loud) your inner knowing, this golden guidance, a specific question for that moment. One such question may be: "What's most interesting and crucial for this moment?" Then wait and see if an answer flashes into your mind. Don't worry if one doesn't. It may take a day or two for a reply to show up. It may even come to you as a comment or word from someone else.

When you trust your body's first response, the guidance you get is high quality. As Goethe said many years ago, "Just trust yourself. Then you will know how to live."[21]

More recently, the late great Steve Jobs (can you tell I admired Steve Jobs?) also said, "Have the courage to follow your heart and intuition. They somehow already know what you truly want to become. Everything else is secondary."[22]

## Why not learn to trust yourself?

I've said this before, but it bears repeating: only you truly know you. You already have what you need inside to guide you. You just have to tap into it regularly. You will get better and better at this. I know this seems a bit airy-fairy. Believe me, I am a science-based individual and probably would have rolled my eyes initially too, but I have seen this work for me. I know there is truth to this intuition concept because I know what I have felt in myself or have seen when things felt wrong or right.

For years I searched in books, listened to audio programs over and over, and attended lecture after lecture hoping to find the answer to what I should do with my life. Remember, I was always looking for the 'easy button' answer from some external source. I wanted someone to come out and tell me, either through written words in a book or verbally said to me. For years, I had been accustomed to others leading my life–teachers, parents, colleagues, and friends–all well-meaning, of course. Then it hit me. I realized that only I can feel what is right for me about things and certain situations. Others cannot feel my emotions. So how would they know if something is right for me or not? Only I can feel that for myself.

You need to live the life *you* want and be guided by *your* emotions. Our lives are made up of moments strung together. If we experience many happy moments, then we feel our life is happy. The opposite, unfortunately, is true as well. With many unhappy, sad moments, we sense our life overall is unhappy. You need to feel happy about your life. You have that right. Don't live your life for others; live it for yourself. Only you have control over yourself. Take a stance and be responsible for your life. Do what excites you! Start today. Do not delay your happiness any longer!

We need to see that our role is to engage with life, to participate, to enjoy, and have fun–just like our animal friends.

"In the end, it may be the quality of our deep inner experiences that counts the most to our souls, not just what we accomplished in the world."[23] Accessing your connection to your body and life energy serves your personal growth, creativity, and well-being. This is why your emotions are so important. Your feelings and emotions are part of your inner experience, and only you can sense them. As Brian Tracy eloquently states in *Focal Point*, "Each of us can decide only for ourselves what makes us happy. And each of us can decide what makes us happy only by listening to our inner voice and following its guidance and direction."[24] So, what you need to do is direct your life toward ways and means that will give you those happy feelings, then you will be living a happier life.

Now that you have learned how to tap into your intuition, I would propose you use a combined, more holistic approach that uses your intuition and rational mind together when making decisions for yourself and about your life. Developing an increasing awareness of your body and learning to integrate body, mind, and spirit connects you to your deep inner knowing and considers all aspects of you. By doing so, it will help reclaim your innate energy, resilience, and guidance.

While the Information Age of the recent past focused on taking in information that answers what and how, the Intuition Age or movement of today focuses on answering 'why.' Even leaders at companies and organizations are leading with this same philosophy–start with why. An entire book has been written about this subject, aptly entitled *Start with Why*, by Simon Sinek. Helping to tap into your reasons why, your reasons for doing something, essentially will have more personal meaning for you.

The trend nowadays is to a more holistic approach; using the mind as well as bodily signals (both truth and anxiety) for truer and more wholesome solutions.

Use this framework for tapping into your intuition, trusting it, as well as considering what you know in your mind to help answer life's questions, to live more in the flow and with more meaning. Eventually, this process will become second nature for you. Give yourself permission to take time with this process. There is no need to rush. Take the time to know yourself. Take time to think, even daily. Ask yourself the hard questions regularly. Invest this time in yourself. The payoff will be huge for you.

Know what you like and what you want to be and do. Assess if you are where you're supposed to be. If not, consider making the necessary changes to get on track. Assess how you feel after making the changes. Jot down your notes and feelings in your personal journal. Track your progress. Have regular check-ins with yourself, at least annually if not more often, to assess if you are on the right track.

Your life is a journey. Don't expect to get it all right in a matter of days. By being in tune with yourself, trusting your intuition, and knowing your strengths and desires and what excites you, you will be well on your way to living a more purposeful and fulfilling life.

So let the process begin.

The next chapter will have you starting to Dig . . . get your proverbial shovel ready!

## FETCH STEPS

"Sit and Stay" to Tap into your Intuition:

1. Set aside fifteen minutes every morning or evening of quiet time for you to be alone. Turn off any electronic devices or background noise.

2. Get in a comfortable position, close your eyes, and clear your head. Do not think about the tasks ahead for you. Only focus on your breathing.

3. Spend five minutes focusing only on your breaths as they go in and out.

4. Once relaxed and with a clear mind (think of your mind as a clean slate or chalk board), ask yourself a question like this one: "What is most crucial for me at this moment?" Wait for an answer. Don't worry if an answer does not "pop" into your head right away. It may take one or two days for it to reveal itself.

5. Repeat the steps above daily. You will get better and more comfortable with this. Get good at tapping into and trusting your intuition. It will help build your self-confidence and your decision-making skills and leave you happier in the process.

# 5

# Dig to Find Your Gift

**Now that you've learned about trusting** the process of learning from animals, as well as trusting your own intuition or what I like to call your "animal instincts," you are ready to dive in and start the DIG process in finding your gift. This chapter is a working chapter. Read it through first, then set aside a few quiet hours to work through the questions and thought processes to get the most out of this chapter. This exercise is **very important** to you. This is where you will discover your Gift.

## DAP and SOAP

The DIG process follows a system similar to what we use in veterinary medicine to diagnose a pet's condition called DAP or SOAP. DAP stands for data, assessment, and plan, while SOAP is the acronym for subjective, objective, assessment, and plan. SOAP is basically DAP one step further, in which the "D" in DAP is sub-divided further into subjective (S) and objective (O). SOAP is what we will be using in your DIG process. We want not only subjective information, which is information about you by you, but also objective data, which is information that comes from others about you.

Not to get too technical here, but merely to show why this analogy is being used to 'diagnose' you and your gift, as a veterinarian, I would collect data such as a thorough history taking, lab data (e.g., blood work, urinalysis), and measurements like body temperature, heart rate, and respiratory rate (often taken in the examination room). This type of data would serve to help narrow down the list of possible diagnoses from which a plan could be devised to either perform more tests or discuss possible treatment options with the pet owner.

Subjective data (S in SOAP) in veterinary practice would be information gathered by the pet owner, like: "Fluffy appears in pain as he has difficulty getting up after sleeping," or "he yelps when I pick him up."

Subjective data can also be what we observe in an examination room, such as fearful or submissive behavior or scratching, itching, or scooting-type behaviors. These are all helpful clues, if you will, to help direct us to a proper diagnosis, as well as helping us to determine how to handle the pet during a physical examination to avoid being bitten (or worse, having them bite the pet owner!)

Objective data, on the other hand, is data that's more concrete such as actual numbers, images or visuals that are measured and/or collected like a body temperature with a rectal or ear thermometer, an X-ray image, or the finding of an ear mite under the microscope from an ear swab of a cat.

When conducting both a good history and a physical examination, being systematic is important so as to not miss an important piece of data. Nothing is more embarrassing than ordering a battery of tests, making an assumptive diagnosis of allergic skin disease, and sending the pet home with hundreds of dollars of oral medications and topical shampoo and conditioners with instructions to the owner to bathe the dog two to three times weekly, only to find out that the groomer found a flea on the dog the next day. How embarrassing!

## Diagnosing YOU

In "diagnosing" YOU to find your gift, we will follow a similar process. We will use the SOAP method, but instead of taking your temperature (phew!) or doing an ear swab on you (ick!), you will be asking others for feedback for the objective data. The subjective data will be answered and collected by you and about you.

Because we can express ourselves more fully than our animal friends, and because we're more complicated beings, I've divided the subjective part of your assessment further, adding more detailed questions into what I call the DIG process. I have incorporated this process into a DIG workbook, which is included in the *Dig. Leap. Play. Companion Workbook* (available for download at companionworkbook.com). You can also just get a blank, lined journal or booklet to write your answers in.

You will want to set aside some quiet alone time and be in an environment where you won't be distracted when thinking about and answering some of these subjective DIG questions. Only you will see these answers, so be honest and truthful with yourself. You will essentially be peeling back the onion layers to reveal the bud of the 'true you.' Listen to your heart and your gut. You may even need to meditate on some of the answers.

The first part of the DIG exercise asks:

## What are your strengths?

Strengths are those attributes about you that are easy for you to do. List all of your strengths that you feel you have. Don't ask anyone (yet) for their opinion. This is the subjective part. What do you feel you are strong in? Write all the strengths that come to mind here in the section below marked "My Strengths." If you need more paper than what's provided here or in the Companion Workbook (at companionworkbook.com), continue writing in the notes section at the back of either book. Keep writing and keep thinking. If you're on a roll, don't stop! If you can't think of any more then stop. You can always come back and add more at another time or try again tomorrow.

### MY STRENGTHS

_____

_____

_____

_____

_____

_____

Now for your weaknesses. Again, don't hold back.

## What do you feel are your weaknesses?

These are the areas where you don't feel particularly strong. It could be that you don't like doing these types of activities, or you have difficulty doing these things, perhaps compared to others. Now I'm not expecting you to put down 'brain surgery' as a task you can't do, but instead, more 'run-of-the-mill' type of tasks or features about you that you identify as weaknesses. Include perhaps a personality trait that you feel could slow you down in progressing toward your goal or prevent you from becoming your true self.

Remember, I'm not expecting you to be perfect here. It's okay to be vulnerable and truthful in writing down your weaknesses. If you only have one weakness that you identify, that is fine. If you have ten or more, that is fine, too.

### MY WEAKNESSES

_____

_____

_____

_____

_____

_____

This is just a subjective assessment of where you are right now in life. It's a snapshot of who you are today. You are getting real. You are meeting yourself, your true self, today.

Even if you've done a process like this before, maybe twenty to thirty years ago, do you think you may have changed in some areas? Maybe you have new strengths that you didn't realize you had or have developed over the past two to three decades? This is why we're doing this process, this deep dive into discovering yourself **again**! I believe this is an important process to do every year or two to ensure you are on the right track for living to your full potential.

You see, this is not about **simply reinventing who you've been, it's discovering who you've become**.

Read that again. Part of the goal of this exercise is to make you realize that you are more than you think. You do have a lot to offer the world. (I told you I was going to help change your attitude; your attitude about yourself!)

Did you complete the "Weakness" section? Good. Now I'll share mine with you. I didn't want to put words in your head before you did it for yourself.

My weaknesses are perfectionism and procrastination. Now I know you may be saying to yourself, "Teresa, since when is perfectionism a weakness?" Let me tell you. When you have this, it's like a double-edged sword. Yes, it can be a good trait and handy at times. For instance, when painting a bedroom wall, being a perfectionist is advantageous allowing no paint to go "outside the lines." However, when you're suturing up an old overweight dog after a tedious spay surgery and subcutaneous fat is causing your suture line to bulge and not end up perfectly smooth, it can be *quite* stressful, aggravating, and can cause you to beat yourself up mentally. More often than not, I find that my need to strive to be perfect in almost everything I do wastes a lot of precious time and energy and can lead to stressful situations and negative viewpoints about myself–all of which are not a good thing. So yes, in many ways, perfectionism can be detrimental to your psyche and self-beliefs, your ability to try things, and thus your personal growth.

Procrastination ties in nicely with my perfectionism because in striving to be perfect, I don't let go of a project until I feel it's almost, or as close to perfect as I can make it. Again, this is wasting valuable time. As I grow older, I'm realizing now, finally, that it's more important to get it done, rather than wasting precious time trying to get it perfect. The 'not getting it done' part is what I fear I will regret later in life because I am so focused on being perfect, instead of just getting it done and out there.

This book is a good example. I started writing this book over ten years ago, but because I didn't have 10,000 followers, as was recommended by a speaker at a Toronto workshop in order for publishers or literary agents to consider taking you on as a client, I got discouraged and stopped. I didn't finish it. But as time went on, and the pull to get my message out to help others who may be struggling and feeling lost and unfulfilled like I did, became stronger (likely due to my intuition here). I cared less about being perfect and worked on getting this book done–once and for all! So, with you reading this book, this is proof that weaknesses can be overcome. (I still tried to make it as perfect as possible, mind you, but I'm super stoked that it's finally out and helping others.)

Okay enough about me . . . for now.

Next let's visit your passions. This is a fun topic list!

## What are your passions?

A passion is defined as a compelling emotion or feeling. This is what gives our lives "the juice." List all the things that you're passionate about. Think about what you like to do on the weekends for fun and enjoyment. What do you envision doing when your holidays are approaching? What did you enjoy doing as a kid for fun? How do or did you spend your spare time and your summer holidays? What makes you happy and excited to do? What puts a smile on your face when you envision doing this?

**MY PASSIONS**

_____

_____

_____

_____

_____

_____

A passion doesn't have to be something you *do* necessarily. Maybe it's a cause that you feel strongly about. Maybe you are passionate about caring for children, pets, or the elderly. Maybe you're passionate about creating an invention to help save time or improve people's health. Again, spend some time thinking about what makes your heart sing and puts a fire in your belly. Think back upon your life and the special moments when you felt elated, lifted, and in your power. Think about times when others may have commented that you were "in your element" doing this. My mom, on a few occasions, had commented to me that I was in my element when I was teaching a training class to a group of adults. And she was right. After eight hours of teaching that class, I got so juiced by their positive feedback and seeing hope and confidence build in them during the day that I could have done another eight hours! Clearly, I'm passionate about inspiring others to be better versions of themselves.

You may not have yet done your passion, but you're still passionate about doing or being it. For instance, the passion to get healthy and fit, or the passion to write that song or book that is burning in you to get out. Write that down here. I hope that the exercises in this book will propel you toward fulfilling that passion. Keep writing, whatever comes to mind. Don't judge your answers—just write them down. And don't NOT write your passions because you're thinking more about the "hows" right now. Just write what's in your heart.

Now let's visit your values.

## What are your values?

Values are those things that you won't compromise on. Values can include things like family, health, freedom of choice, control, or nature. What's important to you? What do you hold near and dear to your heart? Your values may or may not have changed over the years; often they don't. Don't underestimate the power of your values. This is often what drives us and is part of our why. It's behind why we do what we do. List what you value now in this section.

## MY VALUES

_____

_____

_____

_____

_____

_____

And no personal assessment is complete without including a hobbies and interests' section. I joke here a bit, but by no means am I downplaying the importance of this section. In fact, on the contrary, I believe your hobbies and interests hold a lot of information and clues about your potential gifts. Hobbies are what you do for enjoyment and perhaps for money whether on a part-time or full-time basis. Think back to when you were a young girl or boy. What did you do for fun in your spare time?

I can recall loving writing poetry while sitting on the back porch of our family home during the summer months as a young girl. I once wrote a poem about my mom for a Mother's Day contest on the radio–and won! I also used to love to sketch with pencil or charcoal, also on the back porch. (I guess the back porch was a good environment for which my creativity exuded. I loved being in nature and still do! Our yard was at the bottom of an escarpment–full of trees and greenery). Perhaps that grounded feeling from nature allows my creative juices to flow. This too is good information for me to know.

I have always loved to read, since as far back as I can remember. I used to love to get lost in reading a book. I can recall while on a family vacation in Northern Ontario, reading an entire book by Danielle Steele called *Wanderlust* when I was about sixteen years of age, while lying down inside a canoe that was tied to the dock (not the best for one's posture, but a quiet, peaceful setting in nature, once again). Reading has been a lifelong passion of mine. I read almost every morning to get ideas, expand my mind, and grow. I tend to read self-help books now, rather than novels, and have since the age of nineteen, but occasionally I like to venture into fiction as an escape. Writing, on the other hand, was something I started in high school but never considered it as a career until later in life.

In grade thirteen (I'm dating myself here), our high school allowed students who maintained an average of eighty-five or higher (as a percentage out of 100) in grade twelve the opportunity to choose to do a creative writing class rather than having to take the Shakespeare and English literature classes that was part of the grade thirteen school curriculum. Fortunately for me, I was a stellar student and was thrilled to do the independent creative writing as my English lessons for grade thirteen. This was my first foray into writing. In this class, I started writing my first book called *The Flowering Weed*. I loved the process, but unfortunately never finished it because (here come the excuses . . .) high school got busy with getting prepped for university, and I was working at least two or three part-time jobs and taking piano lessons. And of course, I had to find time to socialize and go dancing with my friends! But this process sure was enjoyable. I got lost and absorbed into it, and I certainly enjoyed the creative process. And who knows? Maybe I'll pick it up again and complete it (could be yet another 'Procrastination Project' completed . . . stay tuned!)

It obviously wasn't a strong enough passion at the end of high school, because I went into the sciences and veterinary medicine in university instead. However, I am proof that it's never too late to pick up a passion again at a different stage in life.

By looking back into my past, in doing my own "DIG," I can recall many instances of where I liked to teach. I remember as a little girl, standing beside a Little Tikes easel in front of my dolls and stuffed animals, pretending to be a teacher. Then as I got older, I enjoyed helping many of my classmates both in public and high school with subjects like spelling, calculus, and Latin, and then later in life, teaching and leading my employees and business colleagues in the business world.

From all my life experiences to date, with the knowledge, skills, and analogies that I have gathered from different industries and roles that

I have been in, I feel my role in life now is to write, teach, and inspire others to find and use their gifts to help others. And in so doing, not only feel more joy, passion, and fulfillment in their lives today, but to also have no regrets later in life when it may be too late to shine that gift (for both them and me).

It's only been over the last decade or so that I have been feeling more pulled to want to help people through my written and spoken messages. Helping people find fulfillment and joy from the unique perspective, learnings, and experiences I have had in veterinary practice, then taking the leap into running different businesses and getting into an extraordinary level of fitness in mid-life has been personally fulfilling. So yes, reviewing your many hobbies and interests, even at this stage of your life, does matter!

Write down the past and current hobbies and interests that you enjoy doing.

### MY HOBBIES & INTERESTS

_____

_____

_____

_____

_____

_____

Sometimes ideas resonate differently when at a different place or stage in your life, like writing is for me now. What was important twenty years ago may be less important now and vice versa. What didn't seem as important twenty or thirty years ago can be very important to you now. Perhaps your kids are on their own and you find yourself with more time and freedom on your hands. The idea of painting or baking for fun or as a career may be more appealing, even exciting, to you now than it was years ago! Perhaps you love to travel and the idea of taking a group of mid-lifers on an excursion or bucket list trip is right up your alley, now that you can. Think big. This is your life. It's time to take the wheel and direct it in the direction that will most excite you. Use your gift, build on your strengths and desires, and take action in using it to ultimately make you feel alive, useful, and joyful!

Up to this point, the sections in this DIG exercise have been about you over the years, some of which hasn't changed all that much. This has truly been an exercise about you looking back over your life, maybe seeing some of the same patterns repeat because they are ingrained in you now. Perhaps you've experienced some aha moments that you hadn't noticed before because you were so busy just living your life. I hope you found some great nuggets, nevertheless.

Now we're going to look at your beliefs about yourself, both positive and negative. Self-beliefs are the guiding principles and assessments we make about our personal capabilities and what outcomes we expect because of our efforts. These are opinions and judgments that you hold about yourself and often can be what holds you back. These are good to assess so that you can be aware of what are you saying to yourself, about yourself–good and bad.

What are some of the positive, more supportive beliefs that you are holding right now about yourself?

### POSITIVE OR SUPPORTIVE BELIEFS ABOUT MYSELF

_____

_____

_____

_____

_____

_____

What are some of the negative or limiting beliefs you hold about yourself?

### NEGATIVE OR LIMITING BELIEFS ABOUT MYSELF

_____

_____

_____

_____

_____

_____

Beliefs will come into play later when we set goals for ourselves and create an action plan. But they're included here because it's a very important part of the subjective assessment of you; assessing how you think about yourself, your abilities, and capabilities.

Look over your answers about your strengths, passions, values, hobbies, and interests. Are you seeing any trends or patterns? Are some traits repeated in more than one section? Do you like what you see? Is it a surprise to you, or not?

## Trends or patterns to note:

### TRENDS OR PATTERNS ABOUT MYSELF

_____

_____

_____

_____

_____

_____

So that completes the subjective part of assessing you.

Next is the objective part of your assessment, in which we will be asking others for their feedback about you. You will be asking three to five people you know, who know you well, what they see as your strengths, weaknesses, and passions. These people could be your spouse, parent(s), sibling(s), best friend, or business colleagues. Try to get a good cross-section of people from different areas of your life (for example, personal and career) and ideally of different sexes and ages to gain a variety of feedback.

To provide more direction to your family and peers, here is a sample of an introduction letter and some potential questions that you can provide either verbally, in an online survey, or in an e-mail:

I have provided a link to this letter here: drteresawoolard.com/family-and-friends-letter in which you may edit as you wish, and sign before sending it to your family and friends.

Hello friend/family member,

I am working on a self-assessment project that involves doing a thorough inventory check on myself, in searching for my true gift(s) and I could use your help. I am looking for some feedback from people who know me well, and I thought of you.

Would you be so kind as to answer the following questions about me, in terms of my strengths, weaknesseses, and passions? Please do not sugarcoat your answers. I am looking for honest and truthful feedback. This means a lot to me, and I thank you in advance for your time and attention in providing this valuable feedback about me.

1. What would you say are my strengths? In other words, where or how do I shine more than others in terms of either my abilities, skills, and/or personality?

_____

_____

_____

2. What do you think is holding me back from realizing my full potential? What are those things I tend to struggle with, don't enjoy doing, or could be better at? Perhaps, I may have mentioned this to you in the past and I simply need a reminder here.

_____

_____

_____

3. What do you believe are my passions? How would you describe them? Where have you seen me at my best and what was I doing?

_____

_____

_____

Thank you kindly,

Signed by You.


Beliefs, hobbies, and interests are not part of the objective data collection process because they are more personal to you and thus were included only in the subjective part.

Once you've gathered the responses (i.e., the objective data), list all the strengths responses (from your family, peers, etc.) together, then the weaknesses responses together, and then the passion responses together. Compare them to your responses. Are you seeing patterns emerge? Are the responses from your peers and family consistent with your responses? Are there some surprises that you didn't expect? If you need or would like more information, or are curious about certain responses, then call them up and ask them to explain it further. This could be golden feedback for you that may set you on a path of enlightenment!

Sometimes with our strengths or gift(s), it's so natural for us to use them that we overlook these as commonplace and assume everyone has them or can do the same. Often, it's not until we ask others for feedback that our true strengths or gifts are revealed. We often don't give ourselves credit for what we can do and that we do better than others. We downplay our gifts because we don't want to appear conceited or as a 'show-off.' I want to change your perspective about this notion and how you are currently handling your strengths or your gifts. You need to recognize and ideally use your strengths or gifts more often, because this is what you are good at! The world needs more leaders, and by you recognizing and shining your gift, you can potentially help so many others who wouldn't get to benefit or improve if you kept it all buried like a bone.

In this assessment of the responses from both you and others (i.e., the "A" part of the SOAP process), have you been able to 'see' your gift? Has it now become obvious? If so, then see it visually and clearly by writing your gift statement here. A sample gift statement (mine, actually) follows:

## A Sample Gift Statement:

*My gift is to use my communication skills of writing, speaking, and editing to help and inspire adults to find, use, and share their gift from the perspective and ease of learning from animals.*

### MY GIFT STATEMENT

_____

_____

_____

_____

_____

I hope you now have a really good sense about yourself and have found your 'gift' through this DIG process. Use the results of this DIG process as your guiding principle in moving forward in life. From now on, try to choose to do things that will incorporate the use of your gift. The next chapter will have you thinking big about what to do with your gift. Keep an open mind and dream again.

## FETCH STEPS

1. Go back and ensure that you have written in all the sections (e.g. Strengths, Weaknesses, Passions, etc.) in this Dig Chapter. Take this time, once and for all, to truly go inside and dig for the answers. I want you to *feel* the answers. If you prefer not to write in this book, then write them in the *Dig. Leap. Play. Companion Workbook* available at companionworkbook.com, or in a blank journal, that you can easily refer to.

2. Memorize your gift statement. Make it part of your story about who you are. Tell it to others when you meet them or at a networking event. Your gift is what you bring to the table for the feast of life. Practice saying it out loud. Learn to *bark*, or tell others. The more you say it, the more you will hear and believe in it for yourself. You will gradually build more confidence in yourself and your gift. By opening up and being more vocal about expressing your gift, you may be amazed at what opportunities can start opening up for you to share your gift. Embrace it and have fun with this!

# 6

## Stretch Like a Dachshund

**Assuming that you have found your gift now**, how good would *you feel* if you could be doing and sharing your special gift(s) on a regular basis, seeing it help others, and making an impact? Are you using your gift in some capacity now? If so, great! Could you do more with it than you are currently doing? If you're not using your gift or strengths in any capacity, why not? What's holding you back?

Could you create a goal for yourself in which you are using your gift or using it to a greater degree than you are currently? Could you make it a "Stretch" goal–one in which you will be stretched to achieve it? A Stretch goal could be likened to a BHAG–a big, hairy, audacious goal. This is a goal that takes you outside of your comfort zone, stretching you to achieve more than you think is possible. Whatever you call it, this goal should excite you. This type of goal, in and of itself, should push you to do what is necessary for you to grow. I know it's scary to venture into new territory or do something different, but it can also be exciting, maybe even life-changing. You just need a plan, some resources and tools, and the right attitude!

You'll be creating an action plan in the next section, what I call your Leap Plan, and we will be highlighting the resources you need in the pages to come, but first, I want you to just dream a bit here for me. You need to share your gift(s), to let others see and enjoy it. The world just may be waiting for you.

Write a Stretch goal that you would like to achieve by using your gift.

**MY STRETCH GOAL OF USING MY GIFT**

_____

_____

_____

_____

_____

_____

Now, let's fine-tune this Stretch goal and make it into a SMART goal. A **SMART** goal is a goal that is **specific, measurable, attainable, relevant,** and with a **timeframe**.

## Specific

Be specific about what you want to accomplish. The more specific your goal, the better. Clarity is golden here. When you're clear on your goal and you can envision doing it or having it, your subconscious brain works on finding opportunities and making connections for you. You'll be more alerted to people or activities that could help you along the way to achieve your goal.

Be specific about your goal. Who, if anyone, needs to be included? Why is this a goal?

_____

_____

_____

_____

_____

_____

## Measurable

A measurable goal is helpful, but more as a benchmark. It has an end point, so you know you've achieved it when you reach the endpoint.

How are you going to measure your progress and the successful accomplishment of your goal?

_____

_____

_____

_____

_____

_____

## Attainable

You want a goal that you can attain so it sets you up for success; success in becoming more of the person you were meant to be. You don't want it so easy that you won't experience any transformation in the process of achieving your gift, but on the other hand, you don't want to set an impossible goal, either. Aim to stretch yourself a bit out of your comfort zone, because then you will get to experience more of what you're capable of.

Do you have the skills/tools required to achieve the goal? If not, what would it take to accomplish them? This goal should be driven by you and your efforts, not relying on others to do the work. What is the motivation for this goal?

_____

_____

_____

_____

_____

_____

## Relevant

Your goal should be relevant to you and in alignment with your gift, passions, and values.

Is your goal in alignment with your strengths, skills, interests, and values? If not, it should be. Why are setting this goal now?

_____

_____

_____

_____

_____

_____

## Timeframe

Over what timeframe have you allotted for yourself to achieve this goal? Setting a timeframe creates a sense of urgency so you will focus on it. Otherwise, it becomes easy to say things like, "I'll get to it later," or, "when I have time, I'll do 'X,'" and we all know how that goes. We rarely seem to get to it. If this goal to use your gift, or to use it more fully, is important to you (and it should be for more joy and fulfillment in your life and is why you're reading this book and doing the exercises), then make time for it. Make it a priority and be intentional about achieving it.

What is the target date for achieving this goal? Is it realistic and will it create a sense of urgency for you?

_____

_____

_____

_____

_____

_____

Let me stop here for a second and check in with you. I want to ensure that I haven't left some of you in the dust or buried in the dirt trying to dig for your bone. If the initial DIG process, which included the subjective tasks as well as the objective tasks in gaining feedback from your peers and family, were successful in helping you to finally find your gift and you are prepared to move forward in creating a goal in how to use your gift, then congratulations and skip to Chapter 7.

For those who are still struggling to find their gift, even after doing the DIG process, then let's stop and recap here and try something else.

You've completed the DIG process–diving deep into yourself and answering questions about your strengths, weaknesses, passions, hobbies, and interests. You've asked others for their feedback about the same. In coming down to the assessment part of the SOAP process, it'll be the common themes or patterns that you observe that will give you clues about your gift.

Did an idea or description, whether from you or another person, come up that excites you? Perhaps that's a clue to explore further? Is it becoming more obvious what your gift may be? Did you discover that you have more than one gift? And yes, that is possible and quite common. If that's the case, is one stronger than the other? Which one excites you? Which one do you see yourself 'running with?' Could it be something you do for a career or to earn income, whether full-time or part-time? That would be great, but remember that money is not the goal of this book here. It's about helping you to find your gift and it's in the using it, and even better, the sharing of it, that will bring you joy and fulfillment. You've heard it before that "that money does not buy you happiness." However, earning income from using and sharing your gift would be like adding a cherry on top of your gift sundae. It would be like killing two birds with one stone (a bad analogy, granted, for a vet) but you get the picture. It would be like getting two rewards for the same action–using and earning from your gift!

If you're struggling to find your gift even after the DIG exercises, then perhaps doing a deeper dive into finding out more about yourself using another method would be helpful here. Just like in veterinary practice, preliminary tests and history aren't always enough to make a diagnosis. Often, we need further tests to delve deeper into finding a diagnosis.

You may need to do the same here. It's not that you're more complicated, but perhaps you need additional methods to reveal your gift. Further 'tests' for you may entail doing additional exercises like the tried-and-true Myers-Brigg Test, the Clifton Strength Finders, or the Kolbe Index. I did them all and found that they each offered additional insight into simply knowing myself.

The Clifton Strength Finders 2.0 profile that is part of the *Strength Finders* book series by Tom Rath reveals your top strengths. The top five strengths for me were more descriptors of how I do things, and less so of what I should be doing or like to do. I would also recommend taking the Kolbe-A Index. This is an online assessment that measures the instinctive ways that you take action. It was quite revealing and spot-on for me. While there is a fee for this assessment, I believe that the more we know about ourselves, the better we can direct our actions and lives to give us more joy and fulfillment. In addition, reading books like *Your Unique Ability* by Catherine Nomura and Julia Waller, as well as regular journaling, can both be insightful and revealing as to what you tend to like, enjoy doing, and what and where your strengths lie.

You could also take the survey exercise that you did with your peers and family members one step further and conduct a live focus group. Hearing people's comments and bouncing ideas off each other can be both invigorating and fun. Brainstorming or masterminding with others with the sole focus of helping you would be a productive and insightful exercise for you to do, anyway.

Are you smiling when you see the image of yourself using your gift? If so, that's a clue! Work with that for now. Remember this:

> *"You don't have to be great to start;*
> *but you do have to start to become great."*
> – ZIG ZIGLAR[25]

Now, if after conducting some of the above-mentioned exercises you have discovered your gift, go back and write your Gift Statement and Stretch goal that involves using this gift.

To recap, you should now have a Gift Statement as well as your Stretch goal that states what you want to do with your gift. You will know what positive and negative beliefs you currently have and how you can shape these to help use your gift. Finally, you have some idea of some of the rewards that could help drive you and which you will be including in your Leap Plan.

You are doing great and are well on your way to creating your Leap Plan. But first, we need to ensure you have the proper Fetch mindset; one that will support you in moving forward, because your body won't move if your mind doesn't instruct it to. That is what the next chapter will focus on.

## FETCH STEPS

I mentioned at the beginning of the introduction to the DIG process that you may need to meditate about some of these questions in searching for the answers. If you haven't done that yet, then that would be a good exercise to do, as well. While we used others for feedback in the objective part of the assessment of you, only you truly know yourself deep down. This is why meditating and sitting with yourself is helpful.

## MORE SIT AND STAY EXERCISES

1. Spend time with yourself daily–think about or journal your answers to the following questions. Take 10 to 15 minutes every morning or evening to get calm, focused and then ask yourself the questions below:

   i. What would you be doing if you could?

   ii. What would you love to do?

   iii. What do you think you could offer to others that would help them and make you feel good in the process? Don't worry about the hows just yet.

   iv. What do you envision for yourself?

   v. What words or images flash in your mind when you ask yourself, "What is my gift?"

2. If necessary, rewrite your Gift Statement.

   _____

   _____

   _____

   _____

   _____

3. If necessary, rewrite your Stretch Goal as a SMART goal.

   _____

   _____

   _____

   _____

   _____

# 7

# Be Pawsitive

**Before creating your Leap Plan**, you need to develop a Fetch mindset. What's a Fetch mindset? A Fetch mindset is one that is crystal clear on what you want and where you want to go, is focused and determined in getting it, and takes action on achieving it. Think 'bulldog' here. Nothing will stand in your way of achieving your goal. A Fetch mindset builds on a foundation of a positive mindset and adds an action component.

The definition of mindset is an established set of attitudes and beliefs held by someone.[26] A Fetch mindset takes it one step further, focusing on beliefs and attitudes that will propel you to take action in the direction of your goal. This chapter will focus on getting you to think positively while the next chapter will have you build on that to create a Fetch mindset, dramatically increasing your chances of achieving your goal(s).

## A Pawsitive Mindset

A positive mindset steers the body in the right direction. The body gets going based on what your mind tells it. A positive mind directs the body in the right direction whereas a negative mindset can send it in the wrong direction. Think of your mind like a steering wheel, directing your body where to turn, what to do, and how to do it. We need it in synch and in proper gear first, before starting any action. In order to accurately direct your feet to take action in your Leap Plan, your mind needs to be primed and ready to know how and what to say. It's about saying the right things in your head that will encourage you to take the steps necessary to accomplish your goal. This is why having a positive mindset, one that

72

is optimistic and focused on possibility, is so important. Being able to tell yourself things like, "I can do this," or, "I am ready to live the life that I deserve," serve as internal helpers, if you will, encouraging you to do what you need to do to accomplish your goal. It's also about what you feed your mind that matters.

Some of you may already have a positive mindset. Congratulations! Having a positive mindset can go a long way in assisting you in achieving your goal(s). I encourage you to still read the suggestions from the list below, and if an idea strikes you, then start incorporating it into your daily habits. There's probably no such thing as an over-positive mindset anyway, and besides, the more powerfully positive, the better. It's also very attractive simply as a personal trait, and we all could use more positivity in the world.

If you don't feel you have a positive mindset now, incorporating some or all of these methods here into your daily routine can put you on the path to creating one. It's a matter of creating regular habits over time that will become part of your routine. Then, all of a sudden, you are thinking positively naturally. Start with one or two ideas and then add more as you conquer the first few, so as to not overwhelm yourself. The more suggestions you apply from the list below, the quicker and better will be your mindset in terms of being positive and supportive of you.

## Creating a "Pawsitive" Mindset

To create or intensify a pawsitive mindset, you need to properly feed and fuel it. Just like we need to eat food to fuel our body (and hopefully make good choices here too), there's a variety of good choices we can make to do our minds good. The following suggestions will produce better results if done a bit every day as opposed to a lot just once a month. Again, it's about creating the right daily habits that will eventually lead you to your goals and to being the person you were meant to be.

## Methods to help create a Pawsitive Mindset:

- Positive Affirmations
- Meditation and Mindfulness
- An Attitude of Gratitude
- Personal Development
- Body Movement
- Proper Nutrition
- Social Connection
- Belief from Others

## Pawsitive Affirmations

Creating pawsitive affirmations (i.e., short phrases or statements affirming what you want) and then saying them repeatedly to yourself on a daily basis (ideally), whether in your head or out loud, will work wonders for you in creating a pawsitive mindset, priming your mind and body for readiness, and building your confidence.

Examples of pawsitive affirmations could be something like, "My book will impact thousands of people and attract new opportunities my way," or "I can do this!" Affirmations could also be action-focused, such as, "I'm enjoying doing my daily habits that will get me to my goal." These affirmations are beneficial because they work to get you to expect positive results as well as propel you to take the necessary action steps to get those results. Nothing happens unless you take action.

## Meditation and Mindfulness

Meditation and mindfulness are practices aimed at helping to clear one's mind and focus on being more present. It's about listening and trusting your subconscious mind as well as your body's intuition.

For those who haven't tried meditation yet, here are a few keys:

1. Find a quiet, private spot away from external distractions.
2. Sit or lie comfortably. You may even want to invest in a meditation chair or cushion.
3. Close your eyes.
4. Make no effort to control your breath; simply breathe naturally.
5. Focus your attention on your breath and how the body moves with each inhalation and exhalation.

By focusing your mind on your breath, your mind remains clear, like a clean slate. After several breaths, ask your mind a question like, "What should I focus on today?" The answer may simply pop into your mind right then or even later that day. You just have to be aware and wait for it. I've meditated often, and while I'm not an expert by any means, it has helped me to get clear on whatever issue I am seeking answers to. So, give it a try. Be sure you are framing the questions positively, however. For example, asking something like, "Which project will bring me the best outcome?" rather than "What do I do if I fail?" The latter question, of course, is focused on failing, and your subconscious brain and you are hearing the word, "fail." Avoid such negative words and concepts. Focus instead on pawsitive and encouraging questions, words, and phrases.

## Attitude of Gratitude

Life is far from perfect. I think we can all agree on that. Things happen in our lives that can cause us to react negatively, like not making the sale you were hoping for or not getting the appreciation from your children like you expect. We are human and somewhat instinctive beings (at least I am). Having a pity party when these negative events happen is okay once in a while. Just don't live there. Sure, it can serve as an emotional release to cry, but do it and be done with it. Feeling sorry for yourself, or living in the past, does no good for your confidence and state of mind now, nor for your progress moving forward. As Taylor Swift would say, "Shake it Off," and move on.

Having an attitude of gratitude forces you to get outside of your problems and look at the bigger picture. In turn, you are better able to bounce forward when challenges occur in life. Having an attitude of gratitude is also about being more aware of your surroundings and choosing to be grateful for them. Noticing the beauty of nature around you, the people in your life, and the opportunities available to you can put you in a positive state of awareness that is both encouraging and delightful.

## Steps to developing an Attitude of Gratitude:

1. Make it a conscious habit daily.
2. Express appreciation for your life and all that you have in it.
3. Look at the bigger picture.
4. Take the focus off of yourself.

Having an attitude of gratitude is regularly expressing appreciation for your life and all that you have. It's being intentional about noticing what is around you and being thankful for it and the people in your life. Make having an attitude of gratitude a habit by doing so daily.[27] It will set a positive tone for your day. By being observant of all that is around you (nature, your loved ones, your possessions, etc.), you take the focus off yourself and learn to focus on the present and the positive. You realize all the good that is around you and start being more open to greater possibilities. You'll feel more grounded and appreciative in the process, and not fall into the trap of taking things like natural beauty, a place to live, food on the table, and people for granted.

## Personal Development:

Like I've stated before, getting on the path of personal development from an early age has helped me immensely become the strong, confident, capable person I am today. I strongly urge you to make this a daily practice. It could do wonders for you. Personal development includes activities like reading or listening to a self-help book or audio book, attending a seminar led by an expert or motivational speaker, taking courses or classes to expand your mind or your body, and even activities like journaling. Any and all of these activities can improve your sense of self-confidence and give you the motivation to take action on your goals.

## Reading or Listening to a Self-Help Book

Get started by reading ten pages of a personal development book or listening to a self-help audiobook daily. Such activity could spark a business idea, fill your head and heart with warm and positive thoughts to start your day, or even reveal an inspiring story that resonates with and fuels you. My days start early in the morning (usually around 5:00 or 6:00 a.m.) and always by reading a self-help book while drinking a glass of water and a cup of coffee. I can't think of a better way to start the day! Often, I will read something that re-affirms my path in life, makes me think, or gets the creative juices in my brain flowing, having been sparked by an idea in a book. Encourage your kids and loved ones to do the same (maybe not getting up at 5:00 a.m. but at least to embrace the habit of personal development). Build your personal development library as your legacy to then pass on for future generations to enjoy and benefit from.

## Journaling as a Part of Personal Development

Journaling is an important part of personal development because it involves writing your thoughts and ideas about yourself, your goals, and your aspirations down on paper or in a journal. In essence, it creates a visual picture (in words and perhaps diagrams) of yourself. When you see words on a page describing your likes, interests, and desires, often it's easier to see the bigger picture about yourself. It's like having a bird's eye view of YOU. Many of us are visual, and when we can visualize our likes and interests and where they may intersect or complement each other, ideas can more easily pop in our heads, or such observations can make it easier to "see" answers or ideas to explore.

Sometimes when you've read or heard something in a self-help book, a thought gets ignited. Rather than rely on your brain to remember these thoughts or ideas, jot them down in a journal. Write these thoughts and ideas down as you read your self-help book. Use your journal as a companion guide alongside the book you are reading, or ideally, use the *Dig. Leap. Play. Companion Workbook* for this book—found at companionworkbook.com. You can always come back to it later in life, but at least you haven't lost it. Use the back section of a journal or the Notes and Comments section of the Companion Workbook to jot down ideas that come to mind. Reading them periodically, or later, when you are at a different time and place in your life, may spark something in you to do something with that idea. Often, it's timing and/or perspective that can make or break an idea or affect what you may do with it.

## Body Movement

Is it any wonder why dogs go crazy to go for a walk? Maybe they know something we don't, like the numerous benefits to our physical health and mental well-being. Nothing clears the head better than going for a walk or run, ideally in nature, taking in the heady smells of the forest ground and trees and feeling the warmth of the sun. It's truly cleansing for the soul and mind and feels good knowing that you are moving your body and staying flexible. Let your mind wander, say an affirmation out loud, show gratitude, or listen to a motivating podcast while walking or running. Rejoice in the beauty that nature offers, and you'll start your day with a fresh pawsitive attitude every day.

Some of my best ideas and thinking came about while on a walk or run along a wooded trail. I read that Steve Jobs often held walking meetings with others to think better and more clearly. It's as if movement gets the heart beating faster and perhaps pumping more blood to the brain, allowing for better thinking.

Taking fitness classes, running, or walking regularly, and incorporating weight training into your regime, in my opinion, are especially helpful in boosting not only one's physical strength, but mental and emotional strength, too. Feeling strong and fit makes one feel powerful, confident, and almost invincible. In fact, I like thinking, feeling, and reciting this affirmation as I run because I totally believe it:

## *"Strong body, strong mind."*

It's a great feeling and one that I would highly recommend that you start today. You will be amazed at the results.

## Proper Nutrition

Supplying your brain and body with the nutrition it needs for proper function will make a huge difference in how you think, the thoughts you generate, and the energy your body has to take action. Having sufficient carbohydrates, fat, protein, and water, and in the right balance, will enable you to work adequately and synergistically.

Food and mood go hand in hand. Nutrition plays a big part in helping set a more pawsitive mind.

First rule: don't skip meals! Eat regularly to provide consistent calories and to prevent sugar and mood swings.

Eat foods high in protein like oats, grains, legumes, beans, and seeds. Eat a variety of fruits and vegetables to get the necessary minerals, vitamins, and fiber, and eat good fats like avocados, nuts, and cheese, which are good for your brain. And be sure to drink lots of water to hydrate for purposes of flushing toxins from your body, preventing dehydration, and keeping your skin supple and youthful looking.

## Social Connection

Connecting with others is very helpful in creating a pawsitive frame of mind. We are social beings and need social interaction to feel connected, share ideas, thoughts, stories, and emotions, and to give our souls a lift.

Laughter, especially when shared with another, can do wonders in creating a good mood and pawsitive mindset. Surrounding yourself with positive, happy people who care about and support you will instantly boost your morale and make you feel good about yourself. Call up a friend and book a coffee or lunch date. Incorporate these important connections into your planner. It's these moments that will lead to fond memories, keep life light and grounded, build connections, and who knows, may even lead to a business idea or other fun life adventures. Seize them. This is what life is about.

## Building Belief from Others

Seeing or hearing another human being who has accomplished a goal like yours can boost your confidence and belief that your goal can be accomplished. Think about Roger Banister who, in 1954, was the first human to break the four-minute mile. Once he did, many accomplished the same because they saw and believed it could be done. Oftentimes, we need to believe something can be done before we engage in activities to achieve a goal.

By working on your internal voice through affirmations and feeding your mind pawsitive ideas and thoughts, you create an internal environment of belief. By attending seminars by leaders or people who have accomplished great goals, you can also build on your pawsitive mindset of "what is pawssible." So, get out and attend motivational lectures–you never know what nugget or idea may spark in your mind!

## Other Tips

Some other tips to create a pawsitive mindset are to remind yourself of your Why–the reason your goal is important to you to accomplish now in your life. Perhaps make this one of your daily affirmations. Read your Why to yourself daily to serve as a reminder of why you are on this path. Hopefully, if it's strong enough, it alone will encourage you to stay the course.

Also, re-reading your positive and supportive beliefs about yourself could support and build your confidence about yourself. Saying them out loud to yourself with emphasis and conviction in your voice and allowing you to hear yourself will "trick" your brain into truly believing and accepting them as truth.

Sometimes, just taking a break or having a nap to rest your brain (and body) is all you may need to keep it in the right, positive space, as well as ensure you are getting the required seven to eight hours of sleep each night. A tired mind and body may struggle to stay positive and motivated, when what it really needs is rest. So, heed your body's signals and give it what it needs first.

You need to pull out whatever is in your arsenal to keep yourself motivated and encouraged while going through your Leap Plan. Not because it's difficult, but because humans are notorious for wanting to give up too soon along the journey, thinking they are not capable of, nor worthy of, achieving their goals, or that can't accomplish them. But you won't be doing that here, right?

Don't be overwhelmed by this list of suggestions in creating a pawsitive mindset. These are just suggestions, but I do encourage you to try a few of them. They can do amazing things for you, and having a pawsitive mindset is truly going to help put you in the right frame of mind for believing you can accomplish not only the goal you set out for yourself in your Leap Plan (shortly), but anything you go after in life. You will also be amazed at how others will be attracted to your "pawsitive" nature!

The next chapter will have you add an extra layer onto your now pawsitive mindset to morph it into a Fetch mindset, which will be even more beneficial for you. Get ready to . . . Go!

## FETCH STEPS

1. Do at least one of the pawsitive mindset suggestions outlined in this chapter daily for a week.

2. Add a second suggestion on to your daily routine for the following week.

3. Journal how you are feeling now that you have started building your pawsitive mindset.

# 8

# Create a Fetch Mindset

**Now that you understand the importance of**, and hopefully have or are working on building, a pawsitive mindset, it's time to create a Fetch mindset. To reiterate, a Fetch mindset is focused on taking action; in particular, taking action on your goal(s). In my opinion, and from years of observing people around me, what I believe is missing for many who fail to attain their dreams or goals is a lack of taking action. This was the biggest difference that I saw between animals and people. Animals simply act, whereas humans tend to overthink things, talk themselves out of doing something, build up the excuses to justify their inactions, and simply don't take the action they need to do in order to succeed. If this sounds like you, and it's deemed to be due to a lack of belief or confidence in yourself, then hopefully by instilling some of the daily rituals laid out for you in the previous chapter for building a pawsitive mindset, you can overcome this hurdle. If it's from a lack of energy, then read on for some tips to help in building up your energy reserve. If it's from a lack of ability or time, then leverage others to do those tasks that you cannot do or don't enjoy doing to free up valuable time for yourself or learn to say "no" to tasks that are not in alignment with your goals and values. More on this later in this chapter. By learning from animals to start taking more action on those things that align with your gift, goals, and values, you will be more successful in accomplishing your goals and succeeding in life.

To summarize the difference between a pawsitive mindset and a Fetch mindset:

- A pawsitive mindset is one that is in your mind and holds beliefs about yourself that you can do something.

- A Fetch mindset, on the other hand, is about taking action. It takes a pawsitive mindset one step further, ensuring "the something" gets done. Because . . . nothing happens without taking action.

## Purpose

The purpose of having and creating a Fetch mindset in this book is to get you to take action on the necessary steps that you will lay out in your Leap Plan to get to your goal.

A Fetch mindset is also focused on filtering out, and not putting attention on, the many distractions that can easily derail you from taking action toward your goals. Distractions such as internet surfing, TV watching, playing video games, etc.–can take a lot of your spare time away from your own goals (and life) and make you less productive at the end of the week.

## Judgment Time

In developing a Fetch mindset, you have to stay focused on your goal and not be swayed away from the important steps in your Leap Plan. Realize, however, that there will be other tasks you need to do in life. This is when you need to be focused but also flexible. Many things are out of your control, and at times you will have to make judgments about how you spend your time. Get in the habit of asking yourself, "Is this task or use of my time in alignment with my goal(s) or values?" If not, and it's not that important to you, then don't allow yourself to get distracted. Instead of getting caught up in the whirlwind of menial tasks, prioritize what needs to get done based on the importance to you. Otherwise, you'll end up wasting valuable time and energy and delaying the end-point for your goal. If you keep letting distractions derail you, causing you to focus on other people's goals and plans, you may find it difficult to get to your own, thus losing momentum and perhaps wanting to "throw in the towel" and give up. I don't want that to happen for you, so please don't fall into that trap.

## Daily Process

In creating a Fetch mindset, just like when training animals, it's best to make your focused activities a daily habit. Activities like visualizing and taking small baby steps toward your goal. You will learn more about baby steps in the chapter on Preparing to Leap. For now, adding the act of visualization to your meditation and affirmations process will help in creating a Fetch mindset.

Invest twenty minutes of quiet time daily, ideally the first twenty minutes upon waking (or twenty minutes at bedtime before going to sleep) to meditate and to do affirmations and visualization. Pick the time that works best for your schedule, whether morning or evening. The key is consistency and for you to make it a habit as much as possible.

The first fifteen minutes of this routine could be spent meditating, clearing and calming your mind, and focusing on your breath only. Once your mind is relaxed and cleared, and you've said your affirmations (see the previous chapter), now visualize in your mind the goal you want. Visualize yourself having already achieved it and celebrating the big win for yourself. Visualize even achieving the small wins, the daily activities that you set out for yourself in your Leap Plan. Feel the emotions you are feeling. Experience the joy and happiness. Think about some of the positive or affirmative body signals you will be feeling when doing so. Will you feel your heart wanting to pop out of your chest from exhilaration? Are you exuding an overwhelming sense of accomplishment and joy with a smile as wide as your face? Are you hearing your helpers and pack cheering wildly for you as you finish that race or finally holding your published book in your hands? See yourself being surrounded by your loved ones, family, friends, colleagues, and maybe even your clients–all cheering you on. Feel the love all around you and absorb it. Take it all in because you've "earned" this.

With daily focus on your goal (and mini wins), and consciously visualizing it happening for you, you will prime your subconscious mind for the day, alerting you to potential opportunities that may come your way to help you succeed. When you visualize before going to sleep, your subconscious mind goes to work for you while you sleep, coming up with gems of ideas in your dreams or just before you wake up. (It's best to have a pad of paper by your bedside to jot down ideas as they come into your mind, whether you wake during the night or first thing in the morning, just so you don't forget them).

John Assaraf, author of *The Answer*, recommends doing this process of calming, affirmations, and visualizations midday, as well, to keep you focused and on track throughout the remainder of the day. Do what you can, but the key is to be consistent. [28] The more you practice, the more primed your mind will become. As more of your thoughts start resonating with the universe, you will begin attracting what you need to achieve your goal. Opportunities and resources can start to appear quickly or may take some time. Either way, be aware and open for them and be patient.

## Fuel Statement

As was mentioned in the introduction to this chapter, many people fall into the habit of overthinking what needs to be done, sometimes creating the hurdles and reasons in their mind why they can't or shouldn't do something. With this type of "negative" mindset, people may choose to not even bother trying, and thus lose out on the exhilaration of achieving their goal, along with the personal growth and learning that comes from trying. Of course, it's easier and less work to make excuses than it is to do the activity. Or telling yourself that you will get to it tomorrow. We all know that tomorrow never comes. These have become like "knee-jerk" reactions for many people, because they lack the pawsitive mindset (that you have now) and confidence to believe in themselves, along with the belief that by taking action, good outcomes can happen for them. Hopefully, you can see why we are spending so much time on building your positive Fetch mindset here. Many of these "bad habits" are either because of a poor belief system or a lack of prioritization on one's own goals. By building on your belief system and changing it to one that will be more supportive, positive, and encouraging for you, and creating a Leap Plan that will keep you focused on your daily tasks and habits, can you see how we can change these habits and knee-jerk reactions to be more productive for you?

If, however, at any time going through this process you think you still need a boost to take action, then create an energy or fuel statement for yourself that gets you charged up and motivated, like "1, 2, 3 . . . Go!" On "Go," get up and do that thing that you want or need to do. A fuel statement that gets you charged up removes the ability of your brain to rationalize and potentially leads to procrastination. If it's putting on your running shoes to go for a walk, then do it. Here's a perfect example of how the mind directs the body. If you say the right things to yourself that are in alignment with your goals, and you heed the commands, then chances are very good that you will succeed. Think of a dog being released from his leash in an open park and told to "fetch it" as the owner throws a ball. Do you think the dog will wait, sit down, and think about it, or go after it immediately upon being told to fetch it? Right, so stop overthinking and just do it. You'll feel good about yourself for doing it, too.

## Utilize Your Resources

To achieve your goal(s), you will not only need a Fetch mindset that is action-oriented, but also one that helps you be more mentally and emotionally disciplined. You will also need to utilize as many resources as you reasonably can in order to do the daily disciplines required to achieve your goal. Resources like determination (more on this in the chapter on Dogged Determination), a pawsitive attitude, your attention, energy (through body movement and nutrition), and time. In addition to these inner resources, utilizing external resources like your pack or accountability partners will help immensely here too to keep you on track and hold you accountable. If you do, you will have a much better chance of succeeding compared to if you didn't call upon many or any of these resources. If you don't rely on your external and internal resources, then you will just be relying on willpower, which you will soon see is not successful most of the time.

## Energy as a Powerful Resource

I mention energy in the list of resources above from which to draw, yet I haven't touched on this other than mentioning the importance of proper nutrition and hydration. Your level of energy is a powerful resource that is needed in order to take action of any kind. The more energy you have, the easier it is to do a task, whether it be exercise, introducing yourself to potential business owners, or speaking in front of potential investors. Your internal energy level helps to put a pep in your step and intonation, excitement, and passion in your voice. Do whatever you can to help yourself build up your energy reserve. As was mentioned in the previous chapter, proper nutrition and body movement will go a long way toward fueling you and providing the energy you need. Sleep is another important component of energy.

## Sleep as an Energy Source

It is recommended that adults get at least seven to eight hours of sleep nightly for optimal health. If not, we may suffer from a lack of energy.

A lack of sleep has many deleterious effects, such as:

- Negatively affecting our daily activities
- Causing an inability to focus
- Procrastinating on tasks
- Lacking motivation to participate in exercise or activities
- Being irritable or "edgy"

If you suffer from poor sleep, here are some recommendations that may help to improve your sleeping patterns and ability:

1. **Exercise**

   A 2011 study in the *Mental Health and Physical Activity Journal* found that adults aged eighteen to eighty-five who did 150 minutes of moderate to vigorous exercise per week had a 65 percent improvement in sleep quality and reported feeling less tired.[29]

2. **Meditation:**

   Focusing on your breath and calming your mind can help you sleep peacefully.

3. **Writing and Journaling**

   Writing out your to-do list at night can relieve your mind of having to remember the tasks for the next day. Journaling your thoughts and feelings from the day can help get it off of your chest and mind and can also help you remember details and memories that you want to remember.

4. **Reading Before Bed**

   If you don't want to read, listening to an audiobook can take you into a "pretend world" where you can escape and calm your mind at the same time, readying you for a peaceful, dreamy sleep.

5. **Diet**

   A healthy diet full of good fiber from whole fruits and vegetables, along with avoiding sugar, caffeine, and/or a large meal before bed, can help put your body in the proper state for a good night's sleep.

6. **Aromatherapy**

Using aromatic plant extracts for therapeutic purposes has been used for millennia. Calming scents like lavender, Roman chamomile, and neroli in essential oil diffusers can be used just before bedtime to create the right environment for sleeping.

7. **White Noise**

Some people enjoy and need white noise, like a circulating fan in motion, to help them fall asleep. (I personally need no noise to fall asleep, but each person is different and has their own needs and habits to help them fall asleep). There are now Apps you can get to mimic white noise and other colored noises. Try and do what works for you.

If after trying some or all of these suggestions, you are still having trouble sleeping, invest in a sleep app or seek a healthcare professional who can do a sleep study to assess what type of sleep stage is troubling you, and perhaps why.[30]

## Time as a Resource

Another important resource you have is time. You may not think so, but you do. You just have to learn how to better manage and prioritize it. Everyone has the same amount of time each day, yet many ahead of us have done some amazing things with the same amount of time. You can too! Luckily there are now lots of tools and resources for you to use to help free up some time for you and keep you on track and avoid wasting valuable time. Tools like setting an alarm on your phone to alert you on when and for how long you can spend doing a particular activity. You can use a color block system on your computer calendar that blocks out and sets boundaries on certain activities to focus on a particular day and time.

There are many automated software tools that you can use, like AI (artificial intelligence) or CRM (client relationship management) programs that will create or send automated messages out on your behalf to clients and/or employees, saving you time from doing it manually. You can also leverage others to do work that needs to get done both at work and around the house, such as a virtual assistant, house cleaner, handyman, and bookkeeper. If it is not financially feasible to hire such help, then perhaps call upon a family member, neighbor, or friend to give you a hand with some of these tasks. You can always return the favor to them at a later date when they may need you, or it could be more fun sharing the task and doing it together. Ambitious high school or college students, who are interested in earning extra income, could be helpful here too, in doing some computer work, for instance, filing, housecleaning or odd jobs around the house. This will free up valuable time for you to focus on doing what you need or want to do that is in alignment with your tasks and goals. By minimizing the many tasks on your to-do list and staying focused on what you need to do to achieve your goal(s), you will greatly increase your chance of succeeding at accomplishing your goal(s).

You can also learn to simply say, "No." Saying no to things that are not in alignment with your goals and values can also free up valuable time for you. Learn to politely back out of making commitments to everyone who asks for your help, otherwise, you have little to no time to work on your own commitments. Remember, we are not here to please everyone, but to find, use, and share our gift to better the world. And to do so requires us to stay focused and on task to get it done.

## Getting Off Track

Even in utilizing the resources you have, you will still get off track in sticking to your Leap Plan and either miss some tasks or days. It is not the end of the world. Simply recognize it as such and get back on the horse. Follow your Leap Plan and pick up where you left off. If you need to adjust your plan, then go ahead and adjust the time, task, and/ or end-point goal. This is the benefit of tracking the tasks that you do accomplish so you can assess and make changes sooner rather than later, saving valuable time in the length of the timeframe you have allotted for finishing your goal. This is your Leap Plan. Use it as you wish to make it work for you. I am simply giving you some ideas and best practices to help increase your success rate and to make it simple and fun!

In summary, I hope you try some or all of these suggestions, ideally incorporating them into your daily routine to help foster a positive Fetch mindset. You'll be surprised how far this alone will take you. If you develop a pawsitive belief system, attitude, and a Fetch mindset, and take the actions necessary, then anything is possible for you. And that is exactly how you want to position yourself to succeed.

## Now go use that Fetch mindset.

Create new ideas for yourself, write them in your journal, put a plan in place (i.e., a Leap Plan), and set your intentions for taking action. Then get ready to enjoy the ride of your life, because you are going to make your life extraordinary! And you'll be thankful that you did. The next chapter is the first in the Leap section, focused on creating the right habits that you will incorporate in your Leap Plan in Chapter 10. Up to this point in the book, you have been focused on getting your mind and belief system primed to take action and believing you can do what you set out to do. The following section is where this preparation phase pays off.

I am excited for you to work on the next section called The Leap. You are ready, my friend.

## FETCH STEPS

1. Start incorporating the act of visualization into your daily routine of building a Fetch mindset. See yourself doing and succeeding at the activities you need to do.

2. Put a pad of paper and a pen by your bedside to capture those ideas that come to you in the middle of the night.

3. Create a fuel statement for yourself that will fire you up when you need to get moving. Write it in your journal or here below and commit it to memory, then use it as needed.

My Fuel Statement:

_____

_____

_____

_____

_____

4. If necessary, rewrite your Stretch Goal as a SMART goal.

_____

_____

_____

_____

_____

5. Write down the things you can do to free up an extra three to four hours per week (and indicate how much time you're saving for each). For example:

| | |
|---|---|
| Less TV watching (1 hour per day during week) | 5 hours |
| Less internet surfing (30 minutes daily) | 3.5 hours |
| Delegating tasks | 2 hours |
| Automating emails or newsletters | 1 hour |
| Saying "no" | 1.5 hours |
| **Total Time Saved** | **13 hours saved per week** |

| Where to Free Up Time: | |
|---|---|
| **Task** | **Time Saved** |
| | |
| | |
| | |
| | |

# LEAP

# 9

# Become a Creature of Habit

**Animals love routine.** As veterinarians, we would recommend that pet owners train their dog or cat at the same time every day. Training time becomes routine, and they anticipate it. If it's made fun (as it should be), such as by using many yummy treats during training time, they actually look forward to it. Many pets also enjoy the one-on-one attention with their owner, and this time together can further strengthen the pet-pet owner relationship.

Children also enjoy routines in their day. We see this fact supported early on when they attend daycare centers and primary school. Routines bring predictability at a time in their lives when there are a lot of changes, many of course that are out of their control, like siblings going off to school, parents leaving for work, or meeting new friends at a playground or school. When routines or daily habits are in place, it gives the child a sense of predictability, even some sense of power or control. When they know what to expect, they are not as worried or fearful. There are no surprises. Habits and routines bring stability to their days and weeks.

The same goes for adults. We also like routine and rely on habits to get us through the many tasks we need to do in a day. Habits like showering and brushing our teeth, for instance. Showering in the morning is part of our routine to get ready for work. We don't have to put it into our calendars; we just do it without thinking. Same with brushing our teeth after a meal. When we were children, our parents would have to remind us to brush our teeth, but soon the act of brushing our teeth became natural for us to do, without being told or reminded.

Did you know that currently 45 percent of our daily behaviors are habits?[31] Obviously, we have become creatures of habit. Why? Because it works. Habits help us to be more productive, more focused, and more accomplished in our day-to-day activities.

## Advantages of Habits

There are lots of advantages to developing habits such as:

- Conserving mental effort to focus on more important things
- Clearing and calming the mind, resulting in less burden on the mind
- Less reliance on memory and willpower
- Less overthinking and commiserating
- Providing a sense of accomplishment
- Creating the means to get more done in a day/week
- Having a ripple effect in creating other habits[32]

Unfortunately, not all habits are good. There are bad habits that can hurt us long term, like poor eating or gambling. Good habits, on the other hand, are very helpful to put in place; habits like walking, reading ten pages of a good book daily, working out, and eating fruits and vegetables. These have been proven to help you grow your mind and to stay fit, lean, flexible, and even extend your life.

Incorporating more good habits can help you on your quest to leading a more fulfilled life, too. Habits that support you in using your gift and striving for your stretch goal, for instance, can assist you in accomplishing your goal(s). This chapter will focus on showing you many ways to create good habits.

When creating your Leap Plan in Chapter 10, you will be breaking down your main goal into small steps or tasks. Making these tasks something you do regularly, even daily, can become habits that you form. And that is a good thing. You want these to become habitual, so you don't have to think about them, or worse, overthink them. You just do them. This is what will make you successful at achieving your goal.

## Do it Daily or Weekly

Sometimes it's easier to create tasks or habits that you do daily or weekly rather than sporadically every few weeks because the latter are easier to forget. I recently trained a small group of business associates and served as their accountability partner. They unanimously chose weekly meetings rather than twice a month because they knew that a weekly habit would be more likely to be remembered than twice monthly. I used to belong to a business networking group called BNI, a global organization with tens of thousands of members. We met at the same time each week for the same length of time (ninety minutes), as does every chapter in BNI globally, because it develops good business habits, encourages strong relationships, and it has been shown to be effective for the past forty-plus years.

In addition to making life easier by creating regular habits, one gets better at doing something when it's done often and regularly. Take my workouts for instance. When I first started my workout program, I felt weak, unsteady, uncoordinated, and easily got out of breath. But after two or three weeks of almost daily workouts, I had progressed to the point where the moves were much easier, smoother, and the weights seemed light. I was starting to see results in my body. It's like the 10,000-hour expert concept first described by Malcolm Gladwell.[33] The more hours you devote to doing art or any task, the better you will get at it. After about 10,000 hours, if done correctly, you could even become an expert at it.

## Have a Strong Reason

Starting a new habit is easy when you know that it will help get you to what you want. Say, for example, you want to lose weight to be able to fit into your dream dress for your son's wedding coming up in four months. You have a strong reason, and you know that by eating healthy and exercising regularly, it will help you get into that dress. Developing good, positive habits in terms of eating right and working out regularly, like daily, will go a long way to get you to your goal of losing weight, and you will likely do it quicker!

## Workout Partner

Having a workout partner is also going to help you stay engaged and motivated in developing the habit(s) that you want. If it's losing weight or getting into shape, having a friend or spouse who is going to join you in your quest will work wonders in that both of you can hold each other accountable. Having a fitness coach would also be beneficial, if not better, because your friend or spouse may be too lenient and let you "off the hook" if you miss a day or eat the wrong foods, whereas your fitness coach can come down harder on you. And that may be a good thing. Coaches are there to help get you to the finish line and be more successful in reaching your goal, not to be your friend necessarily.

## Linking Habits

Linking habits is a good method by which to introduce a new or better habit in place of a bad one.[34] For instance, if you want to start reading ten pages of a good personal development book, start reading when you drink your morning coffee, a habit that's already established. By linking the reading with the drinking of your coffee as part of your morning routine, the reading will become a new habit that you form. Eventually you will automatically do this as part of your morning routine without having to consciously think about it. Think of how many books you could read by the end of the year?!

If you used to have a cigarette with your morning coffee, by having a book in your hand instead of a cigarette, the habit of reading can eventually replace the need or desire to have a cigarette with your coffee. Reading, as the good habit, replaces smoking, which is the bad habit, because reading is now linked to your morning coffee, and not cigarettes.

## Environment-Based Habits

Linking habits can also work against you, however. Think about when you walk into the kitchen at night. Does that automatically make you want to open the pantry cupboard or fridge? You know there's food behind those cupboards and fridge doors. It could mean an immediate reward . . . like a tasty cookie or piece of cake. Many habits are environment-based, so the environment itself sets up habits, both good and bad. My husband has taught himself that to avoid eating at night after dinner, he avoids walking into the kitchen because he knows that the kitchen environment at night can lead to bad habits like over-eating just before bed. He makes sure the kitchen lights are off and the dishes are done (what a good husband, right?) ahead of time, so there is no need (a.k.a. trigger) to go into the kitchen at that "danger time"–just before bed.

Walking into a gym, whether in your basement or at a stand-alone building, and seeing the weight equipment and perhaps a treadmill, is a much more conducive environment to getting you in the habit of working out than walking into a bakery. If walking into a gym facility where you see other people working out hard and seeing their positive results is more encouraging to you than seeing just your equipment in your basement, then join a gym. Do whatever you need to do to help you start the right habit for you. Work with what inspires you.

## Timeframe to Develop a Habit

While many say that you need to do something for twenty-one days in order to have it become a habit, others report it's more like sixty-six days, on average. A 2009 study from the University College London looked into how long new habits took to establish roots and found that it takes anywhere from eighteen to 254 days. But on average, 66 days was the figure.[35] And since then, more research generally points to the sixty-six-day timeframe. In my opinion, if you tell yourself that a task or activity will take twenty-one days to become a habit, then that's what it will take. If you tell yourself it's sixty-six days, then you've got both timeframes covered. It's a self-fulfilling prophecy. If what you are doing is good, positive, and supportive and helps get you to your goal(s), then keep doing it, even beyond sixty-six days! Why stop? If working out and eating clean and healthy is making you feel good about yourself, improving your confidence, and allowing you to feel energetic and flexible, then keep going until you can't anymore!

## Key Habits

Often, when starting to experience the benefits of doing one good habit, another similar good habit ensues. It's like a ripple effect. Starting a workout program, doing cardio and weight training three to five times weekly, and seeing how hard you have to work to burn off 200 calories, you naturally want to eat better and not negate the benefits by eating a donut on your way home from the gym! Working out sets up a ripple effect of also wanting to eat better.

The habit of regularly working out becomes a Key Habit that sets in place other beneficial habits that you will want to begin. Use this concept to your advantage. You could become a great role model for your spouse and kids, too, and start them on the path of eating more healthfully and working out. Before you know it, you all could be walking and working out together as a family and eating more fruits and vegetables instead of French fries and pork chops!

## Habits versus Willpower

Many studies have been done on both habits and willpower. Since willpower relies on conscious thought from the prefrontal cortex of the brain, it uses up brain energy, similar to a muscle that is worked. The problem with relying on this 'muscle' is that it can tire by the end of day, thus allowing natural impulses to take over rather than good intentions. This could be one of the main reasons so many dieters fail at dieting.

They're relying on willpower alone, which can dwindle and lose energy as the day progresses, and they haven't yet learned the power of creating good habits and behaviors that are more automatic and don't require conscious thought. Habits trump willpower every time because willpower is dependent on the state of your mind, which can change.[36] (Hopefully less so now that you have a pawsitive one!)

## Changing Habits

If you have a bad habit and you want to replace it with a good one, decide what the new habit will be. Recognize the trigger(s) for the current bad habit and make a conscious effort to do the new habit instead of the bad one. Do this a few times and give yourself a reward. (Hopefully just doing the new habit is pleasurable for you and you like it. That's always helpful and is like a reward in and of itself.) One example could be that to improve your lunch eating habits, you pack a salad, a small whole wheat bun, and an apple instead of dining at the local greasy spoon. Do it more often during the week, and eventually you won't think twice about eating out. You'll not only be eating and feeling better about yourself, but you'll also save money.

Be patient with yourself. It takes time to develop it into a habit (twenty-one or sixty-six days). Be aware of how good you feel when doing the good habit. Just like my husband with the kitchen at night, initially try to avoid the triggers, environment (for him, the kitchen), or situation that used to be linked to the bad habit. Put yourself in a different environment so the bad habit doesn't happen naturally.

Focus on changing only one or two habits at a time to avoid overwhelm and complicating your life. Make it easy for yourself. Start small and gradually build it up. You will need your conscious brain (prefrontal cortex) and energy at the beginning of changing or starting a new habit, but then as time goes along, the energy needs lessen as a different part of the brain (basal ganglia) is involved when you move to "auto-pilot."

Make small changes. Years ago, my husband and I noticed that we were going through a two-pound bag of sugar each week from adding it to our daily coffee. This was obviously an exorbitant amount of sugar to be ingesting regularly, so we made a decision to stop the sugar in our coffee. But instead of drinking coffee suddenly with no sugar, down from the original two teaspoons each cup, we pared it down to just one teaspoon per cup of coffee. After about two months, we tried it with no sugar. To our surprise, we liked it! Now, we both couldn't imagine having any sugar in our coffees. In fact, someone once gave me a cup of coffee with sugar in it by accident, and I couldn't even drink it. The sugar had changed the taste of coffee to me so much that it was unrecognizable to one without it.

## Waldo as a Real-Life Example

We had a Sheltie named Waldo, who we adopted from family members. He came with some bad behaviors. One of the most annoying habits he had (beyond shoe chewing) was jumping up on people. By repeatedly telling people to turn their bodies and walk away from him when he was about to jump up or when he did, and ignore this bad behavior, he quickly learned that jumping up didn't get him what he wanted, which was attention.

Luckily, he knew some of the basic commands like come, sit, and stay. By giving a command of "come Waldo," then, sit," we interrupted his pattern of jumping up to get attention, and he quickly learned that sit was the act to do in order to get a treat or a pat on the head. Eventually, he stopped jumping up on people. He was a beautiful soul of a dog and one of my favorites throughout my life as a pet owner.

## Immediate Rewards

For both forms of reinforcement, positive or negative, the idea is to do it more immediately rather than providing or delivering the reward later. The brain makes immediate connections between the immediate reward and what caused it. We value immediate rewards because it's a sure thing, whereas delaying a reward for the future is less desirable because it's unsure and far away. The use of immediate rewards is not only a key principle and strategy used to drive behavior change for people, but also for our pets.

By focusing on and rewarding the good behaviors that our pets do, and ignoring the bad behaviors, they learn what to do and what not to do. Take the example of a dog that jumps up on people when they enter the house (like Waldo used to do). Very annoying for people (and potentially dangerous for a child) who come to the door to be almost bowled over by say, a 150-pound dog jumping up at them. The dog may simply be excitable and friendly; however, this is not exhibiting good manners. Our inclination when this happens is to pay attention to him, pet his head or neck (depending on how tall he is), and call out his name. To the dog this is a reward. He is getting immediate attention (positive reinforcement), so he repeats the behavior.

By turning our bodies when he goes to jump up, and ignoring him completely (i.e., no touch, no calling out of his name), he learns that he doesn't get the attention he is seeking by doing so. Instead, if we call his name (to distract him and interrupt his pattern) and give a command like "sit," and he does it, then give him a treat immediately for sitting. He learns that coming and sitting earns the treat.

## Positive and Negative Reinforcement

Markham Heid, author of *The Right Way to Change your Habits*, in an article in the *Power of Habits* magazine, explains that both positive and negative reinforcement can help strengthen a new behavior or habit.[37] Positive reinforcement involves adding a positive reward or treat to encourage a new behavior. For example, if you wanted to follow a whole-food, plant-based diet (something I strongly recommend), and for every day that you did, you gave yourself a green checkmark, then the act of putting that checkmark on your meal plan or calendar then becomes like a reward. This progress monitoring is a positive strategy used by behavioral neuroscientists as a method to help people change behaviors. It works because we like to see progress and it gives us a sense of control, which is an important motivator.

Negative reinforcement involves removing something in response to a behavior. So, with our whole-food, plant-based meal plan example, on days that you "cheat" by having bacon on your Caesar salad, for example, the negative reinforcement is not being able to put a green checkmark on your plan for that day, but instead a red "X." The goal, of course, being to see more green checkmarks on your Leap Plan than red X marks. This in and of itself would be rewarding, especially if you are more of a visual type of person, like me. I would strive to see a lot more green than red on my Leap Plan.

## Social Incentives

You could take this checkmark idea further by sharing it with others, like your spouse, accountability partners, pack, or coach. When others see your progress of continuous green checkmarks, and commend you for such progress, this cheering and support becomes like a second layer of rewards for you. You won't want to let them down by not having it on your plan. We will use this "social incentive" strategy, used by behavioral neuroscientists as a method to drive the mind and behavior in people, when creating your Leap Plan.

As you can see, there are a lot of strategies and benefits to incorporating habits into our daily lives rather than simply relying on our willpower and minds. Bringing more habits into your life is not to make your life rigid and boring, but more productive and easier for you to get to your goals and make your life more fun and fulfilling. Our lives are already 45 percent habit, and you probably didn't even realize that. Let's focus on making more of those habits more productive, conducive, and supportive for your goals and you may find your life more in flow and *finally* going in the right direction for you.

Creating your Leap Plan, which you will do in the next chapter, will help you do just that!

## FETCH STEPS

1. Start thinking about some of the habits you have now that are not in alignment with your goals or your gift, and which top one or two of those habits you would like to change to a more productive one.

2. What habits do you have currently that could become a linking habit to another good habit? Choose one or two and see how you could link a current one to one of those in #1.

3. What rewards would drive you when striving for a big goal? Think of three and write them down. You will be incorporating some or all of these into your Leap Plan.

4. Who do you know that could become a good accountability partner for you who you would feel comfortable in sharing your rewards, progress, and Leap Plan with?

# 10

# Leap without Dipping
# Your Toe in the Water

**Well, you finally made it!** This is the chapter where you start taking action on your goal of using your gift. Maybe you've already used your gift in the past in some capacity, but now you want to stretch it further, perhaps by choosing to make it more of a career for yourself, or stepping outside of your comfort zone to do something brand new while using your gift. Congratulations!

Whatever goal you have set for yourself, now is when you start taking action on it. This is the magical piece, in my opinion. Nothing happens until you take action. I've said this before in this book, but I wanted to say it again because it is so powerful and true.

Up to now, in discovering your gift and setting a Stretch goal, it's mostly been ideas and thoughts that have moved from your head onto paper. Now it's time to move it from your head and paper to your feet!

This is the all-important step in the Dig. Leap. Play. formula. You've spent a lot of time preparing for this important part of the process–the Leap Plan. I call this your Leap Plan because it will take effort to make changes, to change or add new habits, and to leap into this role or chapter of your life. This new role or chapter is the one you create for yourself. You are in charge. I have tried to make it as simple as possible for you, and the previous chapters have already alluded to what is required in your Leap Plan, so there should be no surprises here.

This is a working chapter where you will be creating your own Leap Plan. Have fun with this. This Leap Plan will not only serve as your guide to achieving your goal, but also a tool for life in accomplishing other goals that you set beyond this one. To find the template for your Leap Plan, you can download it for free here at drteresawoolard.com/the-leap-plan or inside the *Dig. Leap. Play. Companion Workbook* (see companionworkbook.com ).

The keys in developing your Leap Plan are to:

1. Use small, "bite-sized" steps on the path to achieving your goal.
2. Choose daily habits that will serve you well in conjunction with your goal and incorporate them into your Leap Plan.
3. Incorporate rewards along the way.

It's that simple.

## Bite-Sized Steps

You'll be breaking down your Stretch or SMART goal into smaller, more manageable steps that you can do weekly or daily. Working with pets, we call this "shaping." We literally can shape a behavior like teaching a dog to do a somersault (as I explained on page 19 in Chapter 2) by instructing and rewarding the many small steps toward a goal.

Similarly with people, it's the little things we do over time; the tiny continuous changes to our daily habits that can result in BIG results. Like B. J. Fogg, author of *Tiny Habits* says, "We must make our daily practice too small to fail."[38] The notion here is that if our daily tasks are small enough, we most likely will do them and be successful at them. So why not use them to our advantage? Here in your Leap plan, you will do just that. You will chunk down your goal into smaller, almost daily tasks that are small and doable.

These smaller "baby steps," or "bite-sized" steps as I like to call them (in keeping with the pet theme), could be what forms into habits over time. By developing regular habits in alignment with your goal, as learned in the chapter about Becoming a Creature of Habit, it will be easier for you to stay focused on the path to achieving your goal. The less you have to consciously think about your big goal and only focus on the daily or weekly bite-sized steps, the greater success you will have. Again, don't overthink it; just do it.

## Hurdle Habits

If you currently have bad habits that you recognize as a hurdle, say surfing the internet for two hours a day, think of a more supportive habit, that is ideally also aligned to your goal, that can replace this "hurdle habit." Give yourself a checkmark on your Leap Plan for NOT surfing the net as well as a checkmark for doing the positive habit, say going for a walk in in nature instead. Ending a bad habit is as good for you as starting new positive ones.

## A Road Map

By going through this Leap Plan, you are creating a road map to follow so you don't have to consciously think about or rely on your willpower and memory to remember what to do every day. (Because we know how successful that is!) Basically, you are putting in the thinking now by planning it out in advance so that once it is created, then all you have to do is follow it. Follow the daily or weekly tasks that you have laid out like a yellow brick road.

Not only will you be writing down the daily tasks or bite-sized steps for your goal, but also the supportive habits, like daily affirmations and visualizations, which will help you be successful. More on this later.

## End Goal

Using the time frame you set out for yourself in your SMART goal exercise, the endpoint will be your End Goal in terms of time on your Leap Plan. Mark this endpoint on your Leap Plan. Perhaps it's a year from now, or three to six months from now. The time that you have allotted for yourself in achieving your goal will dictate how much you break down your goal into bite-sized steps. The more time you have, the smaller the steps (but more of them). So, take this into consideration and do what you think will work better for your situation.

For example, if your goal is to use your gift of writing and the endpoint for completing the manuscript on your first book is six months, then six months from today will be your End Goal. Then you will break down the steps of writing your book over the next six months. Perhaps you will allot four months to writing the first draft and the remaining two months to get it edited. Then decide how many chapters you want to write in each month. For example, if your book is twelve chapters long and you have four months, or 120 days, to write it, then that works out to be one chapter per ten working days. From there, you may set a target of writing say half an hour each day for five days a week to meet this target. By creating this habit of writing almost daily, knowing you have a target number of two chapters to write each month, you will stay focused and on task.

## Helpers

Having your helpers or accountability partners, your Pack, who will be following up with you to ensure you're sticking to your Leap Plan, will also help keep you focused and on task. Be sure to share the results of your Leap Plan with your helpers periodically, like every week or two, so they can cheer you on and give you feedback. Consider hiring a coach, in this case a book coach, because he/she can not only serve as another accountability partner for you, but also guide you in the proper writing, structure, look, and feel of your book. I did so for this book, and I would highly recommend it. I delve deeper into the types of helpers you may want to consider asking to join your "pack" in Chapter 14, entitled Run with Your Pack.

## Major and Minor Rewards

Having learned the benefits of rewards so far in this book, you will also be incorporating major and minor rewards at certain points in your

Leap Plan where applicable. Say, for instance, at the end of each month you set a major reward for yourself like a dinner date with your spouse or the purchase of a nice outfit, which you will do if, and only if, you accomplished your goal of writing two chapters that month.

Mini rewards are smaller rewards like a pedicure or a latte, for instance, for having accomplished half an hour of writing each day for a week. These mini rewards, like my chocolate cake on Fridays during my competition training, are all you need to focus on each week. It keeps it simple, manageable, and fun. Stay focused on and walk the path ahead, not gazing at the large, looming mountain goal in the distance that you want to climb. The latter can appear too daunting. You will get there eventually by continuing to walk the path –one step at a time. Just stay the course and trust the process.

Later, in Chapter Sixteen, I will delve deeper into understanding the importance of rewards, the types you can incorporate into your Leap Plan, and why we need more of them along the way of any goal journey.

## Notes Section of Leap Plan

There is a comments (or notes) section in your Leap Plan where you can make notes on hurdles you faced that day or week, as well as the wins and emotions you experienced while doing the tasks set out in your Leap Plan. These not only are good to report to your helpers/accountability partners as a potential discussion point to work on, but also shows you your progress and development over time. Patterns of hurdles may be revealed, which could contain a clue about how to stop and address them, so you can then consult with your helpers or coach and devise a strategy to overcome or prevent such hurdles in the future so they do not become roadblocks later on.

Just like you learned to be more in tune with and aware of your emotions and feelings by trusting your intuition, what they mean, and how to deal with them, recognizing and addressing these hurdles or roadblocks and finding a solution sooner rather than later can allow you to stay on track with your Leap Plan and your target end goal. It helps you to be in control and proactive rather than reactive and frustrated. Deal with any hurdles as quickly as possible, and use your rewards, whether major or minor ones, to celebrate your wins along the way.

## Key Accountability Person

Designate a key accountability person for yourself. This is likely one of your helpers that you identified in the Dig process, or it could even be someone you connect with in our Dig Leap Play Facebook community who may be on a similar path as you. This key helper or accountability partner will know your Leap Plan and will connect with you weekly or every two weeks (or however often you feel you need) to check in with you. If you rely on yourself alone, it's easy to slack when there's no accountability or repercussions.

Getting encouragement from this person, even a few words of positivity or a reminder of your goal and your why, could be all the fuel you need to get you back on track if you have derailed somewhat in your plan. Knowing that you have to share your progress and results weekly or every other week with this person should keep you motivated.

Tracking your emotions and feelings regularly in your Leap Plan will allow you to see what you have been able to overcome and witness your progress to date (a reward in and of itself).

## Sample Leap Plan

A Leap Plan is a useful tool for any goal you are seeking, not just in using your gift. See a sample of my husband Mark's Leap Plan that he created to help him lose weight, in the Appendix of the *Dig. Leap. Play. Companion Workbook* or on my website: drteresawoolard.com/marks-leap-plan. I was his accountability partner. What he found in going through this process was that the simple act of placing a checkmark in his Leap Plan once a task was completed was in and of itself a reward. He looked forward to checking off his tasks every day. It instantly made him feel accomplished, and checking off the box in his Leap Plan was an immediate reward.

Knowing that each accomplished (and checked) task was getting him closer to his goal excited him and gave him motivation to continue. He accomplished his goal successfully and he said the process was easier than he had expected because he focused on the bite-sized steps (even just the daily checkmarks) rather than the big weight loss goal ahead of him (again, path versus mountain).

While the Leap Plan can be used for any goal, let's keep this focused on your goal for your gift. Here is a sample of a Gift Leap Plan; mine actually for getting this book done: visit drteresawoolard.com/gift-leap-plan.

## Supportive Habits

As mentioned earlier, I will also recommend that you include in your Leap Plan some supportive tasks that will eventually (and hopefully) become habits if you're not doing them already, like affirmations and visualizations. Getting in the habit of meditating daily, saying your positive and uplifting affirmations to yourself, and visualizing yourself already having accomplished your goal and feeling the emotions you expect to feel when that happens will do wonders for setting your mind and attitude right for the day.

By creating such a Fetch mindset, you are prepping and preparing your unconscious mind to look for those opportunities, people, and resources that could assist you in your quest to achieving your goal. Be sure to include these tasks/habits in your Leap Plan, because they will truly help your mind and give you added energy and motivation to do the tasks you have laid out for yourself in your Leap Plan. We are leveraging both the mind and your body here to get you to do whatever you need to do to accomplish your goal and see success.

## The Leap Plan

Now it's time to work your Leap Plan. Get started on following your Leap Plan today! You may not have created the perfect Leap Plan, but that is not the goal. The goal is to take action and start. You can adjust as you go along. You are in control of your Leap Plan. Have fun with this. Share your results with your helpers or your key accountability person, at least. Be sure to reward yourself. Don't skip this part. Rejoice in rewarding yourself–it's proof that you are making progress. As humans, we want and seek progress. Be mindful of how good you feel when you reward yourself and hear positive feedback from others. Take in the information you want to hear and be proud of yourself. I am.

## FETCH STEPS

1. Download the online template for the Leap Plan here: https://drteresawoolard.com/the-leap-plan and follow the instructions. There are three parts to the Leap Plan.

2. Choose your key accountability person and share your goal and Leap Plan with him/her.

3. Include a few supportive tasks/habits in your Leap Plan that will give you a daily boost.

4. Print off your Leap Plan or have it handy in your computer to remind you daily of your tasks.

5. Now 1, 2, 3–Go. Get started on your Leap Plan!

# 11

# Put Blinders on to Eliminate Distractions

**When starting, learning, or working on a task,** especially a new one, it's easy to get distracted. What with notifications on our phones "pinging," emails coming into our inbox, family members, pets, or coworkers vying for our attention, or the list of household duties hounding us regularly, it's no wonder we get distracted. In fact, with technology alone (laptops, tablets, cell phones, apps, etc.), distractions are more prevalent today than they've ever been.

This is by design, of course.

## Social Media Time

With 59.4 percent of the 4.76 billion social media users globally tapping into the top six social media apps (Facebook, X, Instagram, YouTube, WhatsApp, and Messenger) daily, spending on average two and a half hours per day, it's no wonder the app owners are trying to get your attention. These are major marketing platforms eager to get your attention . . . and money![39] [40]

How can we get anything productive done in this technology-advancing world?

This chapter will show you how.

I will share strategies and practices to not only eliminate or reduce the distractions, but also teach you some brain-training exercises to learn how to focus better. Learning to master your mind and, in turn, master your time, will not only free up time on your calendar to do more of the productive tasks that you should be doing, like the steps on the path to your goal(s) that you have laid out in your Leap Plan, but also improve your mental well-being and sense of self-control.

## Focus versus Distractions

While you may think that focus and distraction are opposites, they're really not. Focus is the ability to direct your attention, ignoring everything else, while distraction, on the other hand, is an object that directs one's attention away from something else. To thrive at a task, you need to be able to block out the distractions so your attention is laser-focused on the task at hand. Easier said than done, right? The problem is, like I alluded to in the beginning, there are *a lot* of distractions that we face every day, both external and internal, and both types can be successful at decreasing our ability to focus.

What are internal and external distractions?

Internal distractions are those things we hear in our minds, like our beliefs and thoughts, as well as feel in our bodies, like hunger, pain, stress, worry, fatigue, and discomfort, which can demand our attention and take our focus off of our task. External distractions are other people (family, peers, coworkers), environmental noises (phones ringing, vehicle or construction sounds), or our technological devices (phones, laptops, tablets, app notifications, watches) vying for our attention, which again, can take our attention off of our task.[41] We need to strive to eliminate, or at least reduce, both types of distractions, if at all possible, in order to focus on the task at hand.

## Procrastination as a Distraction

Some distractions are in the form of procrastination, wherein we create replacement activities to supplant or displace the ones that we should be focusing on.[42]

Here are some reasons for procrastinating that serve as a distraction:

1.  Anxiety and fear of the unknown, e.g. of the tasks and/or results of the tasks; fear of success or failure

2.  Perfectionism

3.  Unclear goals (poor clarity around goals or tasks)

4.  Tasks that are boring or not fun (no inherent reward)

5.  Feeling of being overwhelmed with too many tasks, roles, etc.

6.  Tasks that are too complicated

7.  Underlying conditions (e.g. low morale, depression)

8.  Non-conducive environment (e.g. too many distractions)[43]

We've already covered how to improve some of the internal distractions like negative beliefs and thoughts in the previous chapter. Training yourself to say positive affirmations daily as well as having an attitude of gratitude will help counter these mental internal distractions. If you're experiencing physical pain or discomfort, then seek professional help as soon as possible, if you haven't already done so. Hopefully a solution can be found to aid any physical pain that could be distracting to you. If not, perhaps find strategies to help minimize the effects of such pains or discomfort.

To deal with procrastination, utilizing some of the strategies already laid out for you in preparation for your Leap Plan, such as being pawsitive, creating a Fetch mindset, setting a motivating and clear goal in the Dig process, breaking down your goal into manageable bite-sized steps, incorporating fun rewards into your Leap Plan, and having a supportive pack around you will all aid you in countering the urge to procrastinate. This chapter will further assist you here, if procrastination is a problem for you, in showing how and where to eliminate the many distractions that could get in your way.

## Distractions for Horses and Dogs

Just like humans, horses and dogs can also be easily distracted when doing a task. Horses can be easily spooked if something comes into their peripheral vision, which by the way is an astounding 350 degrees, much larger than ours at 180 degrees. To keep them focused on a task such as running in a race or on a street pulling a carriage, horse trainers use blinders (or blinkers) to decrease their field of vision down to about 30 degrees.[44] [45] While this is neither practical nor feasible for humans (although Panasonic did try through a similar concept called Wear Space, a human horse blinder system created in 2015 that obviously wasn't successful),[46] here are some strategies to adopt to eliminate distractions and increase your focus:

1. Turn off notifications and or log out of your phone apps.

2. Set boundaries. For example, batch-check your emails only at certain times of the day and turn off your phone when needing to focus on a task.

3. Control your environment to mitigate distractions. For example, book a study room in the local library or choose a room with minimal distractions (i.e., white walls, a table, a few chairs, perhaps a whiteboard, and a closed door).

4. Avoid multitasking. Focus on one task at a time. The brain can only deal with one task at a time.[47]

5. Be aware of your goals. Refer to your Leap Plan daily and/or use time blocking on your calendar.

6. Utilize your pack and accountability partners to help you stay on track.

7. "Train your brain" to stay focused (see the next section).[48]

## Blinders for People

External distractions are a major concern these days for people, too. Luckily, many distractions are in your control and do not require you to wear a horse blinker (thank goodness!). Why are these a concern? Because they remove so much valuable and productive time from our days and lives.

## Social Media

Take social media alone for example. The average user spends two and a half hours daily on social media.[49] That's equivalent to 864 hours per year, or about thirty-six days. Basically, just over a month per year is spent on social media alone. Imagine what you could do or accomplish if you had an extra month per year? You can . . . if you eliminate social media. Here, I'm talking about the personal social media that you may indulge in, not the social media needed these days for business. If the latter consumes a lot of your precious time, then consider hiring that task out to free up your time.

## Phone Time

Apparently, we also spend (depending on the data source you read) anywhere from three and a quarter to almost five hours on our phones per day, and we check our phones anywhere from fifty-eight to 344 times a day.[50] The latter equates to checking your phone every four minutes! You can see how it would be challenging to focus if you're being interrupted or distracted every four minutes.

## Television and Gaming Time

Add on to this, the time taken to watch television at ten to over thirty-six hours per week[51] or almost five hours per day on average (for those who are fifty-five-plus),[52] or to play video games (for those who indulge) at two to four hours per day, and you can see how much time is truly absorbed in such distractions that could be better spent doing more productive activities, like working on your goals or even your personal growth and fitness level.

## "Not Enough Time"

The number one excuse why people don't follow their dreams or strive to achieve a major goal in their life that could likely bring them more joy and happiness is, you guessed it, "not enough time." Yet 36 percent of users of social media platforms say they do so to "fill their free time."[53] Clearly, there's a disconnect in priorities or a lack of awareness of how much time surfing the net truly devours.

In addition to using up valuable time, excessive use of social media has also been shown to lead to feelings of addiction, anxiety, depression, isolation, and a fear of missing out (FOMO). This perceived obligation to maintain connections and stay updated on social media has been described as "technostress," which can negatively affect one's well-being.

## Beneficial Effects of Abstaining from Social Media

In 2020, a seven-day social media abstinence trial was conducted by Lorna Brown and Darla Kuss of the International Gaming Research Unit and Cyberpsychology Research Group of Nottingham Trent University in the UK. They were studying the effects of social media on various psychosocial factors by asking sixty-one participants, aged twenty to forty-nine years old, to measure their states of well-being, FOMO (fear of missing out), and social connectedness before and after a seven-day abstinence from social media. The study showed that those with higher social media use actively felt more socially isolated than those with lower social media usage. Similarly, a longitudinal research study found that social media users are more likely to report higher levels of loneliness than non-users of social media.

Overall, the study showed that social media usage resulted in a decreased sense of mental well-being while abstinence after seven days improved their mental well-being. Abstinence (or a seven-day digital detox, as it was called in the study) from social media apps also improved social connectedness and decreased FOMO. It was also deemed that by decreasing the cognitive load on the brain from social media, participants were able to focus on other activities.[54]

As you can see, while social media does help to connect us to family and friends, there are also negative and harmful effects that social media can bring to our well-being and emotional health.

Decreasing the time spent on your phone or scrolling through social media will free up valuable time for you to focus on honing and using your gift in the direction of your goal and will likely increase your level of happiness and joy.

Turning off your phone, and even all your devices, if possible, at least temporarily, will not only allow you to do the daily tasks you've set out to accomplish in your Leap Plan, but will also likely improve your state of mind and sense of control over your life.

## Story of Quitting Social Media

Here's a story featured in *Newsweek* in August 2023 about Gianna Biscontini, a photographer, behavioral scientist, and author of *F\*\*\*less: A Guide to Wild, Unencumbered Freedom* who did just that. She stopped social media for several months for mental health benefits, prompted by her dog, believe it or not.

It hit her one day while in a cafe with one of her dogs at her feet looking forlornly at her. She thought, "Oh what a cute picture to post." It was then that she realized the trap she had fallen into. Rather than embracing the beauty and unconditional love expressed by her dog, she saw it as a moment to only capture "likes" on social media. Gianna came to the realization that she posted photos on social media only for validation. She shared in the *Newsweek* article how defeated she would feel after spending hours creating and posting a beautiful photo, only to have gained just a few likes. She eventually realized the wasted time and negative effects to her mental health that social media was causing for her and how we, as a society, have come to accept it. She also realized how that wasted time could have been better spent creating assets or courses, being in nature, or gaining clarity on her business and goals.

Since quitting social media, Gianna has felt more focused, happier, lighter, more productive, and empowered. She has had more time to meditate, play with her dogs, and clean the house. She felt herself "coming back to life" and returning to her hobbies and to the life she had enjoyed before social media in which she was free to be herself and unencumbered.[55]

While completely turning off social media may not be feasible nor desirable for you, consider a digital detox for a few days, or at least unplugging temporarily, to bring you valuable "extra" time, a sense of control, and perhaps improvements to your own emotional well-being.

## Train your Brain Exercises

Sometimes we can't control the distractions in our environment, but we can learn to train our brain to limit the *effects* of distractions on our focus. Here are ten focus exercises that, if done for just two to five minutes per day, can increase your abilities to focus and concentrate . . . at any age:

1. Make notes or a to-do list.
2. Meditate for five minutes.
3. Read a long book.
4. Exercise your body.
5. Practice active listening.
6. Try a counting game.
7. Memorize patterns or flash cards.
8. Do crossword puzzles or sudoku.
9. Visualize an object and recall as many details as possible.
10. Take cold showers.

## Benefits of Brain-Focus Exercises

Training your brain takes practice, but the benefits of these brain-focus exercises are multifold:

1. Improves mental health and self-esteem.
2. Motivates you to set new goals.

3. Halts the habit of multitasking.

4. Helps you to control your emotions.

5. Improves your attention span to be able to learn quicker and more efficiently.

6. Become calmer and more confident.

7. Have greater peace of mind.

8. Improve your attitude and reactions to events and people.

9. Master your mind so your mind doesn't control you.[56] [57]

I suggest just working on one of these at a time until they're mastered or habitual. Even still, increased focus can only be for a set period of time, which can be different for each individual depending on how well they've trained their brain. Sustained attention over a long period of time is tiring and saps your resources. Remember to take breaks and use good nutrition to restore your mental resources and improve task performance. (A good night's sleep is also helpful here to rest your brain and body and regenerate your resources).

And now for more animal analogies and how we can learn from them.

## Making Focus Fun

Finding ways to keep it fun is also helpful to maintaining focus . . . just like it is with dogs! To keep dogs' attention, we use positive rewards like tasty chicken, steak, or liver treats, use fast movements to keep their attention on us, and train in environments initially with minimal distractions. So why not give yourself a "treat?" Take a break and treat yourself to a latte or a nice cup of tea, or a walk around the block for fifteen minutes. Give your mind a break, breathe in some fresh air, and feel re-charged upon your return.

Rewarding yourself for not reaching for your phone every four minutes, staying focused on the daily task at hand, staying physically active, and thus making better use of your valuable time to become the person you are meant to be will be well worth it to you in the long run–to your goals and your life.

Then after you've done this . . . go ahead and post your milestone achievement(s) online for the world to see!

In the next chapter, you will read further about the benefits of creating habits around learning and what it can do for both your brain and your life.

## FETCH STEPS

1. Start the habit of "batching" your emails and only checking them once each in the morning, mid-day, and at the end of the day and assess your productivity after a week.

2. Turn off the notifications on your phone during your workday unless you need them for your work.

3. Turn off your phone at night and do not turn it on until after your morning reading, affirmations, visualizations, and journaling.

4. Pick one of the Brain Focus exercises and try it for a week or two. Assess if you feel you are improving in the area of focus.

# 12

## You Can Teach an Old Person New Tricks

**Having been introduced to the world** of personal development from the young age of nineteen, I learned from my first "mentor," Tony Robbins, the concept of CANI, which stands for constant and never-ending improvement. The Japanese have a single word for constant and never-ending improvement, and it's "KAIZEN," from the Japanese words Kai meaning "change" and Zen meaning "good." In other words, KAIZEN means change is good.[58] The idea here is that while big changes can be good, so can small, incremental changes over time, even daily. By being aware of, and focusing on, making small changes in yourself and your life, no matter how small, you are still likely to improve. So, don't fear change; embrace it. And instead, change your mindset about change.

Think of changes as a means to learn, grow, and improve. I have to say that I have wholeheartedly embraced this concept all my life. Whether it was learning from self-help books, CDs, seminars, podcasts, taking on new skills or roles at work, trying new adventures in my personal or business life, or from conversations and observations from friends, family, and acquaintances, I attribute more of my personal growth, skills and personality over the years from this type of learning than from the formal education I gained from three University degrees. Personal development and this idea of constantly improving myself, have made me into the person I am today–a creator, entrepreneur, and leader whose mission is to help others be the best version of themselves. Thus, I am a strong believer that one should never stop learning. To learn is to grow. And not just as a person, but to also grow your brain!

Yes, your brain.

There is a wealth of research and evidence in neuroscience supporting the fact that our brains can continue to grow even into old age, not only from formal learning but even from doing tasks and solving problems in everyday life.

## The Brain and Neuroplasticity

The concept of being able to stretch the brain's capacity at any age is called neuroplasticity.[59] It does not mean that our brains are made of plastic, but instead that they are actually malleable. Our brains can change structure and function through mental experience. When we learn, our brain changes physically.[60]

What is a mental experience?

A mental experience is any thought, action, or event that can effectively change your brain. The simple process of thinking or learning can trigger neurons in your brain to create pathways and interconnections with one another. Our brain makes connections and new pathways whenever we look for everyday solutions, carry out tasks, or are undergoing deliberate learning.[61]

## The Hippocampus

With more interconnections between neurons, one area of the brain gets bigger: the hippocampus, an S-shaped structure deep in the temporal lobe of each left and right cortices of the brain. The hippocampus is the only area of the brain that can grow new neurons

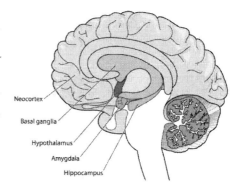

throughout our lives, a process called neurogenesis. The role of the hippocampus is to regulate motivation, emotion, learning, memory, and spatial navigation.[62]

With 86 billion neurons in our brains, each of which can connect to many other neurons, we literally can have about 150 trillion connections or synapses existing in our brains! While the number of neurons does not change, the number of connections does. It is these connections that are responsible for learning and memory. By using newer imaging techniques like functional magnetic resonance imaging (fMRI), researchers have proven that the hippocampus can grow new neurons to about 700 new ones to each left and right cortex every day. In fact, by age sixty, one-third of the neurons in the hippocampus are new, formed through neurogenesis, and that these "new neurons," so far as we know, are used for learning and memory.[63] This is great news, proving that you *can* continue to learn even at an older age!

## Hebbian Learning and Effects on the Brain

Donald Hebb, a Canadian neuroscientist from the 1940s, noticed that rats raised as pets who had more opportunities for social, physical, and sensory stimulation not only performed better at cognitive tasks, but had more extensive interconnections (synapses) than rats raised in laboratory cages. It is assumed that the same processes occur in human brains, but there is as of yet no direct evidence. In the 1980s, studies were done with children raised in orphanages in Romania, where they were deprived of social and physical stimuli. These children had delays in cognitive, language, and social development, revealing that such deprivation of learning is detrimental to their growth.

The process of learning that involves two neurons firing together and forming a connection from a thought, event, or action was called Hebbian Learning (from Donald Hebb) in 1949, wherein Hebb coined the phrase, "neurons that fire together wire together."[64]

Knowing that the hippocampus is required for spatial navigation, a study was done in London in 2006 comparing taxi drivers and bus drivers. Using fMRI, it was shown that licensed London taxi drivers had larger hippocampi compared to the bus drivers. The need by taxi drivers to learn how to navigate around London without a map resulted in an increased firing and connections of the neurons in the hippocampus area of the brain, compared to bus drivers who follow a set route every day. So, for taxi drivers, their hippocampi can increase in size from more interconnections as compared to bus drivers, whose hippocampi didn't change. [65]

## Examples from the Human and Animal World

### ALBERT EINSTEIN

Curious as to what a genius's brain looked like, when Albert Einstein died at the age of seventy-six in 1955, his brain was autopsied and photographed. Years later, the photographs were examined and showed that there were a lot of dense neuron connections in the corpus callosum, the band between the left and right hemispheres of the brain. In fact, when the corpus callosum of Einstein's brain was measured against those his age and to those in 1905, when he was at his most prolific, his was much thicker compared to both groups. This thickened corpus callosum of Einstein, due to more coordinated communication between the two hemispheres of his brain, may be what provided him with his intellectual gifts.

Thus, intelligence or "smartness" may be more about brain connectivity due to learning than from our biology. So no longer can one use the excuse that they weren't born with smarts. Everyone has the same ability, as we all have about 86 billion neurons from birth. You just have to use them to create more connections!

"Intelligence is not fixed, it turns out, nor planted firmly in our brains from birth, but instead is forming and developing throughout our lives." [66] In other words, you can teach an old person new tricks.

## FOUR-FINGER GUITAR PLAYERS

With continued practice, say in learning to play the guitar, the neuronal connections made become stronger through repetition. In addition, these strong connections can form new pathways from the existing strong ones in situations when long-time guitar players lose a finger on their left hand, for instance. The neurons sort of "repair" themselves to replace the missing information from the missing finger. Isn't our body and brain incredible in terms of how it can repair itself?

These strong connections also aid in retaining and recalling memory. When learning similar tasks or concepts, you can tap into already existing pathways, thereby recalling and learning more quickly.[67]

Taking advantage of this concept of neuroplasticity and neurogenesis, research has been done to show that there are ways to enhance the brain's ability to grow new neurons and thus more connections. They are:

1. Exercise—helps prevent neuron loss; 150 minutes of moderate-intensity cardio and a minimum of two days of weight training per week can increase BDNF (Brain-Derived Neurotropic Factor), a protein that stimulates nerve growth

2. Diet—decreased fat or cholesterol and an increase in B vitamins; decreased caloric intake

3. Decrease stress—e.g. mindfulness or meditation

4. Learning—increases connections and new pathways with new learning

5. Playing—e.g. games which can be a type of learning

6. Sleep—to improve growth of the dendrites (finger-like cells present at the end of neurons or nerve cells)[68]

You are probably wondering and asking, "Can lifelong learning help stave off diseases like Alzheimer's?" While that hasn't been proven yet (to my knowledge), a long-term study of 200 senior citizens over seven years has shown that good education combined with lifelong learning does reduce certain degenerative processes in the brain and allows such brains to be better at compensating against any impairments. Education thus has a positive effect on brain aging.[69]

The key message here is to build up those reserves and continue learning until you can't. The benefits of learning aren't just for later in life, nor are they just for the brain. Take learning a new skill, for instance. There are lots of benefits to both the mind and body when learning a new skill. For example:

1. It gives you confidence. Learning a new skill boosts your feeling of self-worth.

2. It helps counter boredom because it gives you purpose.

3. It keeps you healthy. Learning a new skill helps to create and maintain a positive mindset.

4. It motivates you; helps get you up in the morning or out of a rut. It's exciting to learn something new!

5. It boosts your happiness level and lifts your sense of well-being because learning can be fun.[70]

6. It benefits others. A new skill can be applied at work (leading to increased or future opportunities) or at play. You may want to share your newfound knowledge to help others, as either a mentor or leader.

7. You become more interesting, too, when you are more knowledgeable.

There are many stories of adults learning a new skill well into their later years. Consider, for example, the story of Carl Allamby, MD, who became an emergency medicine physician at the age of fifty-one, after years of working as a mechanic.[71]

One of my favorite stories is of Laura Schultz, a sixty-three-year-old grandmother who did an amazing feat out of pure determination when she lifted the back end of a Buick off her grandson's arm. When interviewed by Charlie Garfield, author of *Peak Performance* and *Peak Performers*, she kept resisting any attempts to talk about what she called "the event." She didn't like to talk about the event because it challenged her beliefs about herself, what she was capable of doing, and what she could have been or done in her life. With the learnings from this event and some coaching from Charlie, Laura Schultz went back to school to study geology, got her degree, and taught at the college level–all in her sixties![72]

You may be asking, "What about learning for older dogs?"

Ever heard of the expression "You can't teach an old dog new tricks?" Well, it's false. You can. Just like humans can continue to learn in older life, so can older dogs. Some would argue that older dogs actually learn better in old age because they're less distracted by their environment compared to puppies.[73]

Even though the best learning window for dogs is twelve to sixteen weeks of age (the socialization period) to prevent future bad behavior, dogs can learn at an older age just like people.[74] While learning an entirely new task may take longer for an older dog, a task that's like one they already know will take less time to learn than with a puppy who is learning it for the first time.[75]

## World's Smartest Dog!

Dogs enjoy learning, and training time can be a fun time to bond with their owner. Take the example of Chaser, a female border collie with the largest tested memory of any non-human animal, who from eight weeks of age to the age of fourteen was trained to learn the names of over 1,022 toys. Not only did his trainer's research test the boundaries of the canine mind, but it also showed the immense cognitive capabilities of a dog. Chaser became the world's smartest dog! He could even retrieve a toy he had never seen or been taught to retrieve before by using the process of elimination. We learned from Chaser that dogs are much smarter than we give them credit for.[76]

Betsy, another border collie, had a vocabulary of 340 words and could retrieve objects from just a photo.[77] It is presumed that because border collies are herding dogs, and are highly motivated and focused on their owner's commands, their capability to learn is especially heightened. With proper and repeated training, who knows what limits are possible with our canine friends![78]

## Application to Learning–Importance of Being Creative

When you are learning, let yourself be creative. Have fun with it. Be open and receptive to learning. Keep an open mind and be patient with yourself. It may take a little longer to learn something than you did at age eighteen when your brain was at its peak of learning efficiency, but it's still good news to know that it's never too late to learn.

Lifelong learning satisfies a human need to explore and grow, especially if the learning is in alignment with your goals and aspirations. Learning not only improves the quality of your life, but also your sense of self-worth.

The best ways to learn are by doing, through repetition, by failing or making mistakes, and when emotion is attached to it.

## Learning by Doing

Learning by doing, or what's called experiential or kinesthetic learning, is more effective than just reading and reciting because it's more engaging and rewarding with a hands-on approach. You gain positive feedback such as a compliment or reward by someone, which incites a release of dopamine, making you feel good. This positive emotion also arouses the neurons to get them firing and to build stronger connections.[79] In my experience, learning by doing is my favorite form of learning.

## Learning by Repetition

Learning by repetition helps build muscle memory and, as was just mentioned about long-time guitar players, for instance, neuronal pathways and connections are strengthened through practice, making the learning more cemented and easier to retrieve.[80]

Emotion can be a good teacher, too. Think of the pain we feel when we fail. When we fail at trying something, we may feel pain or disappointment (negative feedback). Such emotions trigger neurons and we learn very well what not to do the next time. When we stop learning, however, the brain "prunes" (or trims) back the neuronal pathways that may have been involved in that process previously, making it difficult to recall or do that task later in life.[81] So, if you don't want to forget a task, keep at it, try a different approach, or learn something new about it.

## Learning by Teaching

One of the best ways to learn something is to teach it, or to learn it *as if* you're going to teach it. If you know you may have to teach something, you'll be more motivated to learn it (like the border collies do). Teaching learned knowledge also builds your communication skills, confidence, and leadership ability.[82]

A Roman philosopher Seneca once said, "While we teach, we learn."[83]

The process of researching, studying, organizing material, and then communicating it is an effective approach to enhancing your learning outcomes.[84]

For me, teaching not only forces me to know my material, but allows me to give back and to share my unique perspective, my learnings, and hopefully inspire someone to take action on a wish or a goal that he or she may have. Knowing this could happen from this book or a seminar that I conduct motivates me and gives me purpose in my life. Even the process of learning itself juices me. When an idea strikes me from something I've heard, read, or observed, and I have then taken action on, especially if it's in alignment with my values and goals, I'm at my happiest because I know I'm growing and making a positive impact on others. We, as humans, have an innate need to grow.

So now that you know that it is never too late to learn a new skill, it is however recommended to do it sooner rather than later to not only build those neuron connections and pathways and not have "pruning" done in the brain, but to keep current in the workforce, protect your future, and stay employed (if that's important to you).

If currently in the workforce, look for opportunities to improve operations or solve problems. Don't be content with the status quo. The creative thinkers are the ones that help a company (and persons) progress. Learn new skills for your career. It will help make you more marketable and valuable for the company. Chances are you will be promoted or retained if the company is hiring for new positions or is restructuring.

Learning skills in other areas, outside the workforce, is valuable too. Additional skills such as public speaking, computer skills, community work, or volunteering may help to set you apart from other candidates for job opportunities or for advancement. Perhaps a committee you volunteered for in your community highlights your leadership and team-building skills, which are required for a new position at your workplace. And finally, those many Toastmasters classes you participated in week after week may pay off and prepare you to apply for a position as a sales trainer at your company.

Gone are the days where you can do one skill all your working life for the same company for forty years and then retire. Most people now change jobs or careers on average five to seven times. With the advances in technology and AI, we all need to try to keep up in some capacity, to stay current and "in the know" or run the risk of getting lost in the dust.

## Hidden Benefit of Learning: Uncovering Talents

Learning a new skill can not only keep you motivated, and is refreshing and enlightening, but also could uncover a hidden talent you didn't know you possessed. This happened to me some years ago. I was invited by my neighbor to attend a direct marketing seminar about identity theft. At the time, I was in the process of setting up a website for my veterinary marketing company and felt this information could prove helpful for my website. So, I agreed to attend. What I hadn't expected was my enthusiasm to want to become part of this movement to help protect others against this growing problem. I was fascinated by this field, liked the business model and the people in it, and saw how I could offer this service to veterinary clinics and their employees as part of my veterinary marketing business. In just fifteen months, I became the number one seller across Canada and was appointed the vice president of group training for Eastern Canada. I was now training groups as large as 300 people and conducting two-day seminars in two different provinces. I was in my element. I learned something about myself through this process. I learned that I loved to teach! This experience opened my eyes to seeing a natural talent I had, but that I hadn't fully realized until this happened. I found I was more energized after conducting a training session for eight hours than when I had started. I was juiced by the feedback I saw and received. Participants were inspired by my trainings. I had found my unique ability. I learned that speaking to groups of people was definitely something I wanted to do. "Is it possible I could become a motivational speaker?" I asked myself back then. While this experience stretched my comfort zone, I had learned a skill that I now wanted to do on a more full-time basis–speaking to inspire others, through animal-inspired concepts such as in this book and future ones. So be open to learning. You never know where it may lead you.

I hope you are encouraged now to continue learning and that I have proven to you that you are never too old to learn. Don't let an older age be your excuse to not learn a new skill, a new language, or follow that creative process you've always wanted to do like writing a song, a book, painting, or traveling.

As you go about using your gift and striving to achieve your goal(s), if there's something you need to do, or need to learn to do, or feel pulled to do, then do it! Because now you know you can. Even just staying active and doing daily tasks (such as your daily habits) will keep your brain in gear, continuing to make new and stronger neuronal pathways and building up those reserves that will likely serve you well into your later years.

Here's to continued learning and to adopting the concept of CANI.

Don't give up on yourself or on your goals. The next chapter will inspire you to keep your nose to the grindstone on the path to your Stretch goal.

## FETCH STEPS

1. Consider enrolling in an online seminar or in-person college course in a field that you want to learn more about. Broaden your skills. Even choosing to upgrade your computer skills, learning about social media marketing for your business or taking a course to enhance your speaking or communication skills, would likely do your business and personal life good.

2. Get in the habit of reading ten pages of a good personal development book every day. You expand your mind, get great ideas that could propel your career or business, become more knowledgeable, and of course, build those neuronal connections to get smarter!

3. If the idea of playing an instrument excites you, then do it. Whether it's taking up piano again, playing guitar, or learning the flute, then go for it and have fun with it. You will not only be able to play music, but you'll be building those brain connections and reserves in the process.

4. If you have an area of expertise that you feel confident in sharing, then offer to teach it at a local college, high school, or adult learning center. It will cement your learning. You'll meet new people, and you will feel good about yourself just from sharing. You may even learn to love it!

# 13

# Have Dogged Determination

**Dogged determination is defined as a tenacity** to continue with, and a refusal to give up on, something even if it becomes difficult or dangerous.[85] While I would never recommend to someone to do something dangerous, having such an attitude or discipline is what truly sets the achievers apart from the dreamers.

When you think of dogged determination, what comes to mind? For me, two images come to mind: one of a bulldog and the other of Winston Churchill. I'm not sure if it's because they are both thick, heavy, and stern looking or because I have been well marketed to over the years by the famous United States Marine Corps logo of its bulldog mascot, which portrays an attitude of toughness and fortitude, but whatever the reason, they both truly represent the term "dogged determination" as I will highlight for you here.

## Bulldogs

The modern bulldog, as described by the Bulldog Club of America, is "gentle, intelligent, affectionate, strong, and determined."[86] Formally bred from a mastiff and a terrier, the bulldog was used for fighting.

Bulldogs are a symbol of dogged determination. They are associated with determination, strength, and courage and will do whatever it takes to get the job done.[87] It's no surprise, then, why the bulldog became the mascot for the United States Marine Corps and one of the most recognizable military logos in the US.[88]

Even the term "bulldog determination" has been used interchangeably and synonymously with dogged determination. Bulldog determination means never giving in to defeat, being creative in finding ways to overcome obstacles, and never giving up, no matter how hard things get. [89]

You can see why having such a bulldog mentality when attacking your daily tasks is important to achieving your goal(s). By locking in on your task at hand, and without other distractions nor a negative mindset clouding your intentions and plans, you will undoubtedly succeed.

## Winston Churchill

You will see shortly that while not a bulldog obviously, Winston Churchill epitomized the character of one through his traits of determination, strength, and courage. Some even coined him the "British Bulldog."[90] Winston Churchill was the Prime Minister of the United Kingdom from 1940-1945, and again from 1951 1955.[91] He fought against Hitler in the Second World War and led his troops from the brink of defeat to victory. He was a great leader who had a willingness to take risks and learn from his failures. Not only was he gifted with insight, but he was also very good at making decisions and forging forward with an iron will despite the many odds that he faced, both on the war grounds and inside himself.

Churchill suffered from melancholy and experienced several bouts of severe depression throughout his life, possibly even manic depression (bipolar disease). Yet through his own dogged determination, resilience, and courage, he was able to conquer his own inner state (which he called, interestingly enough, his "black dog"–a faithful companion that's sometimes out of sight but always returns) as well as the enemies of the country he was defending.[92]

Winston Churchill knew his weaknesses but developed strategies to keep them at bay. He knew that if he was idle for too long, the depressive state would return (i.e. his "black dog" would come home), so he kept himself busy reading, writing, and painting. Having done these activities throughout his life, he ended up being the only leader in history who not only saved a country but wrote the most books (he published over forty books in seventy-two volumes and hundreds of articles)[93] and painted over 500 pictures.[94] Truly a productive strategy, as well.

As Winston Churchill said himself, "*Continuous effort–not strength or intelligence–is the key to unlocking our potential.*"[95] This is why dogged determination is included in this book. It is an important factor in creating your success and helping you to become the person you were meant to be.

What's required to have dogged determination?

Well, luckily, you do not have to look like Winston Churchill or a bulldog to have dogged determination, however there are a few factors that you should have, such as:

1. A strong focus
2. A strong commitment or Why
3. Confidence in your abilities
4. A positive mindset
5. A persistent attitude; one of not giving up[96]

Many of these factors have already been addressed in the previous chapters. In the Dig section, you worked on finding a strong Why, and in the earlier chapters of this Leap section, you learned techniques to help build a positive (pawsitive) mindset, improve your attitude through affirmations, and to focus your brain on singular tasks. Confidence comes from being prepared, taking action, and learning from taking said action. Getting you to take action, learning from the activity, and growing both in confidence and abilities are key goals of the Leap Plan, as well as, of course, getting you to your goal(s).

Your Why is the reason you are committing to taking action. It is the emotional basis of your goal, and the end benefit of taking action on your Leap Plan. For many of you, your Why likely is focused on finding and using your gift, once and for all, to share and better the world in some capacity, and in so doing, find more joy and passion in your life. To give back is to have lived life more fully.

Once you decided to use and share your gift for a worthwhile goal that you set in the Dig section and committed to it (announced it to your accountability partner(s), family, and friends) in your Leap Plan, you ignited the flame. It's discipline, this dogged determination, which is required to fuel the flame; fueling it on the right path and ignoring the many distractions that can come your way to potentially derail you and take you off the path. Remember the chapter on Putting the Blinders on to Eliminate the Distractions?

Making a decision is the first step. To decide (de-cide) in its Latin root means to "cut off."[97] To decide to do something means you cut off any other alternatives from occurring or interfering with you achieving your goal. It is in the act of making that decision, the decision to go after your goal of using your gift, that you really start the commitment process.

A decision (your goal) becomes a commitment once you have cemented in your mind that this is what you want to put your energies into and have resolved to take action on it (your Leap Plan). It's meaningful to you and you want to work on achieving it, otherwise you will regret it later in life for not doing it. Announcing your goal to someone else like a workout partner, friends, or an accountability partner, in other words, your Pack, further cements your commitment to taking the action necessary to achieving your goal. But know this....it is the discipline to continue taking the action (the daily and weekly tasks laid out in your Leap Plan) that will get you to your goal.

Discipline or dogged determination keeps you focused on your goals and tasks. Do whatever you have to do to stay focused. Write your goal(s) on an index card, for example, and read it every morning and night, or even several times throughout the day. Not only will this serve as a reminder for you to stay on track, but you will be more open to opportunities that come your way that could potentially help you with your goal.

There may be times that you feel your efforts are not amounting to much nor seemingly getting you closer to your goal. It's at these moments that I want you to review your Why, your goal, and reach out to your accountability partner to see what could be missing. Is it your mindset? What are you telling yourself? Are you doing your meditations, affirmations, and visualizations daily? Do you need to readjust your tasks or timeframe in your Leap Plan to take some pressure off you? Do you need to try a different approach perhaps, or take a short break? Whatever you do, do not quit on your goal and dream. Keep going. Victory is just around the corner.

## Developing Dogged Determination

Can dogged determination be created or developed?

While some people may have determination as part of their personality, seemingly born with it, I do believe it can be developed through affirmations, a strong belief system, taking consistent action, and seeing positive results from that consistent action. Think of it like a good habit to develop.

Even if you don't believe that determination or discipline is part of your personality, or it's weak, here are some ways to develop it, or build on it, to make it even stronger:

1.  You have to want it badly enough
2.  Narrow your focus
3.  Use daily or weekly accountability
4.  Be creative in your approach
5.  Learn to say "no"[98]

You need to have a strong Why about how important achieving your goal is to you–one that will make you cry. You need to want it badly enough that you will do whatever it takes for you to put the necessary time and effort into achieving it. Your Why should keep you inspired and motivated enough to keep you going until you achieve your goal. By staying focused on only the tasks you have laid out in your Leap Plan and knowing that they will get you to your goal (or at least close enough), you won't run the risk of being spread too thin in terms of energy and time on many other tasks, and thus you can stay focused.

By looking at your Leap Plan daily, which includes your Why, your Stretch goal, and your tasks for each day or week, you can remind yourself each day of what you need to do. Be sure to reward yourself upon completing your tasks–this will also help to keep you motivated. Engage and reach out to your accountability partner often, as this person(s) will hold you responsible for completing your tasks and keep you on track toward your goal. If you're veering off course, or not getting the results you were hoping for, then try a different approach. Experiment, be creative, and make it fun.

By staying focused on your goal and tasks, it should be easier for you to say no to those people or activities that are not in alignment with them. Doing so will free up valuable time and mental energy for you to stay focused on your tasks in your Leap Plan to ultimately reach your goal. Doing so will make you feel more in control of your time and effort, and thus empowered throughout the process. With this attitude and focus, you will surely meet your goal!

To showcase some of these methods in action and to inspire you, I want to share some canine and human examples of dogged determination.

## Canine Examples of Dogged Determination

There are many stories of dogs, beyond just bulldogs, showing acts of persistence. To dogs, failure is not an option. Here are two interesting stories:

Larry Barkan tells a story in the book *Learning Persistence from My Dog* of his dog that tries every day to catch the geckos outside his house but with no success. However, the dog never returns looking dejected or like a failure. The dog does not attach a meaning to "failure." In her mind, she will never fail. She simply hasn't succeeded yet. Isn't that a great attitude? It's certainly one we can all learn from.

Humans, on the other hand, tend to give up too early. We often listen to our self-defeating talk, thinking and talking about failure too readily. We need to take lessons of persistence from our dog friends and not give up so quickly. And we most definitely need to stop the self-defeating talk that keeps us from persisting on the path to achieving our goal(s).

Another interesting example of determination is from the famous

Japanese Akita named Hachiko from Tokyo. From the time he was a pup until the age of two, Hachiko would walk alongside his owner to and from the Shibuya Train Station every morning and afternoon. Unfortunately, his owner suffered a brain hemorrhage and died while at work when Hachiko was just two years of age. The next morning, and for the next ten years, even fighting the pain of arthritis in his older age, Hachiko went to the train station every morning and afternoon precisely when the train was due to come in, waiting for his owner to return. Hachiko sat for hours waiting for his owner who, of course, never appeared. But Hachiko never gave up. His daily activity eventually caught the attention of commuters, train station workers, and a former student of Hachiko's owner who studied Akitas. This student published an article about Hachiko, one of only thirty documented purebred Akitas, and his act of loyalty and determination. The article garnered nationwide fame across Japan, and Hachiko became known as a symbol of loyalty and even likened to a good luck charm. Both a statue and a movie have been made about this amazing dog named Hachiko.[99]

## Human Examples of Dogged Determination

Rudy Ruettiger, author and motivational speaker, became an icon of determination on the football field in the mid-seventies. Overcoming dyslexia, financial hardships, and a short stature (5'6"), he worked hard to get good grades to be accepted into the University of Notre Dame. Working as a groundskeeper at the nearby campus, he kept his lofty goal in sight of playing for the Notre Dame football team. After three rejections, he was finally accepted onto the Notre Dame scout team in 1974. Luckily for him, the head coach allowed walk-in players on occasion to play on the varsity team because he realized that sheer will can often trump experience. Knowing Rudy's determination to play on the field, the coach called upon Rudy to join in the game on November 8, 1975. This was to be his first and last time Rudy would play in an official game, but after tackling the opposing quarterback in the final play, the team went on to a big victory and Rudy became a legend.[100]

Another human example of determination is best-selling author Sherrilyn Kenyon, who stated in an interview with Sara Connell for the *Authority Magazine* that "dogged determination" and being too stubborn to quit were what contributed to her becoming a bestselling writer. She just never gave up. While her first book flopped because she followed what her published friends were doing and writing a "marketable" book, she vowed from then on to write what was in her heart to write. Being poor and having to fight the belittlement and lack of support from her husband for her writing career, often having to write in a closet for hours while he slept, she stayed focused on her lifelong goal of writing. Eventually, by following her heart, she finally got a book deal with a publisher. The next several books became *New York Times* best sellers (she's placed eighty books on the *New York Times* list over the past twenty years), is regularly at the number one spot, and some of her series are being made into a major motion picture or adapted for television.[101]

## A Vital Ingredient to Success

Dogged determination is one of the greatest secrets to goal achievement and one of the most vital ingredients to success. As Peggy McColl stated in her book, *On Being a Dog with a Bone*, "Without determination, dreams will never become a reality."

In history, like in Winston Churchill's case, the people who achieved significant accomplishments did so with a high degree of determination. With determination you create confidence, more determination, faith, and the will to get through any challenge. Without it, people struggle and live a life of quiet desperation. Decide right now NOT to be one of the latter–the quiet desperate ones. Remember, it all begins with a decision. Start exercising that decision-making muscle right now. Go ahead, get mad, get strong. Stand in front of a mirror and say out loud, "I am determined to _____ (you fill in the blank)." Now say it again, but this time with that bulldog look in your face. Say it like you mean it!

How does that make you feel? Did it stir some juices inside of you? Was there a spark inside of you that just ignited? If not, say it again, but this time hit your hand into your fist as you look in the mirror and say it. Feel it, get emotional about it, and believe it. Get yourself fired up. You can do it.

You can do amazing things with decision, commitment, and persistence. Become doggedly determined and do not accept failure as an option. Through continuous effort, like with Rudy and Sherrilyn, you can learn how to get better, improve, and succeed in achieving your goal(s). Hopefully, some or all of these stories, both human and canine, have inspired you to take action with persistence and determination.

So far in this Leap section, you have learned strategies to remove or lessen the interfering distractions that can derail you, learned methods to focus your brain, and in this chapter, learned how to develop your dogged determination. Now, nothing is stopping you from achieving your goal!

So, get after it . . . like a bulldog on a bone.

To make this process easier for you to stay on track and have more fun, I urge you to leverage the many benefits of accountability. In the next chapter, you will learn fascinating facts of how a support system can considerably increase your chance of success in achieving your goal.

## FETCH STEPS

1. Exercise your decision-making muscle; learn to be intentional about making quick decisions. When you are next out at a restaurant, scan the menu and pick an item quickly without changing your mind. Stick to your decision.

2. Read more about the motivating histories of Winston Churchill or Rudy Ruettiger. You will find either story very inspiring.

3. Revisit your Why. Does it still resonate with you? If not, make a stronger one.

# 14

# Run with Your Pack

**In Chapters 8-10, you were asked to find helpers,** those people who can help you when you need an emotional boost or when you run into a hurdle on your journey to your goal. As we all know, in life, things don't always work out as planned. Situations and events can occur outside of your control that may require you to change your plans and find alternative solutions. Doing it alone can feel overwhelming, futile, and exhausting to the point where you feel like giving up; giving up on your goals and your dreams. This is when you need to reach down inside of yourself, remind yourself of your Why, think like Winston Churchill with a dogged determination, muster up your resolve, think positive thoughts, and ask for help.

## Ask for Help

Asking for help should not be viewed as a weakness. Asking for help is a strength because you realize you can't be good or perfect at everything, even though some of us try or would like to think so! This is why you're focusing on your special gift: to find and use your strengths and natural talents to deliver to, and help, the world, or at least your community. By doing so, you also realize that you need to rely on others to help in the areas where you're not as strong or proficient and when you need support, both emotional and practical.

## The Need for a Support System

This chapter is focused on the need for a support system; a pack, if you will, to help you when you "hit that wall," come across a hurdle, or are simply at an emotional low point. After all, we are emotional beings, not rocks. It will also identify the many benefits of having a pack to run to that will potentially expedite your journey and make it easier to achieve your goal. Identifying the different types of helpers who can be part of your pack will also be addressed in this chapter and why you may want many helpers.

We all can benefit from the aid of others both in life and in achieving our goals because we don't have all the answers and because we certainly can't be proficient at everything. Smart, successful people realize this fact. Business owners and entrepreneurs, for example, will often hire those who complement their skill sets or who are better than they are at doing something. Take Henry Ford, for example. He knew that if he hired people smarter than himself, he didn't need to know it all. It worked out successfully for him, wouldn't you agree?

## Benefits of a Support System

A support system can help you thrive. It can provide you with not only better coping skills, but the knowledge and resources you need to solve problems and overcome hurdles. A support system, or your pack, can provide helpful and practical feedback, truthful advice, new connections and friendships, a healthy distraction from the problems and hurdles, reduced stress and anxiety, a sense of control, and even a longer, healthier life.[102] Studies from the Mayo Clinic have shown that people with supportive people in their lives live longer, have better health, and higher well-being.[103]

Social support gives people a feeling of being loved, cared for, respected, and a sense of belonging.[104] Having a social support system can improve overall health, reduce stress and anxiety, and help you stay motivated and empowered.[105]

## Face to Face Is Best

A study was conducted in 2022 assessing the effect of social support on certain stress measurements of twenty-one pairs of long-time close friends placed in stressful situations. The study involved measuring the heart rate, arousal states, and salivary cortisol levels when interacting face to face with their friends and when performing public speaking and arithmetic in front of a panel, the latter two situations of which were the imagined 'stressful situations.' This study was to show how a social support group could affect one's stress levels. The study showed that social support and the ability to interact freely with your friends, even for five minutes, could both blunt the stress response in situations that would normally raise one's stress factors and restore stress levels back to pre-stress baseline levels. The most beneficial factor was the face-to-face interaction between the friends.[106]

## Packs in the Wild

Even in nature we see animals like wolves and wild dogs travel in packs, as a type of support group, to not only overcome stressful situations but to also improve their chances of survival. They hunt and run together to take down game for food for the group and protect each other and their young from other prey—all in an effort to prolong their own survival rate and that of their species.[107]

## Tips for Selecting Your Pack

A support system is defined as a group of people who can provide you with emotional and/or practical support. So, when you find yourself in a tough situation (and it will happen because everyone experiences ups and downs along their journey, especially when you're stretching yourself and entering unknown territory), you will have someone or a group of people that you can lean on for support. It may include family members, friends, and even strangers.

Tips for choosing a good support system are:

1. Someone who knows you well enough to know if something is wrong or if you're hiding from your true feelings and will call you out on it.

2.  Someone you can trust to provide you with honest and truthful feedback without upsetting you or putting you in a worse emotional state.

3.  Someone who will keep you on track with your goals.[108]

You may need different people to provide different kinds of support. For example, your mother, spouse, or a long-term close friend may serve as emotional support, while a stranger that you hire as a business coach, for instance, can provide you with more practical business support, advice, and accountability.

You need to be proactive in seeking and reaching out to your support network. They won't just naturally come to you, unless, of course, they see that you are not able to deal with a situation on your own or are severely stressed or distressed. By being proactive, you can perhaps avoid putting yourself in such situations. Ask family and friends if they would be willing to help you on this journey of yours. Explain to them what your goal is, why it's important to you, and what you may need from them.[109]

## Support or Networking Groups

If you don't have a strong support network in your family or circle of friends or feel you could use more support than what they can provide, then consider joining a support or networking group that is somewhat aligned with your goals. For example, a writer's group, an entrepreneurs' group, Toastmasters, BNI, or your local chamber of commerce. Try one or two and see what appears to be a good fit for you. While there are many online groups in various sectors, I would suggest one that has the members join in person, as I believe relationships can be better created and solidified when people meet in person. And as we just learned from the social support study, face-to-face interaction is key for creating that social support feeling.

## MY WOMEN'S NETWORKING GROUP

When I opened my first brick-and-mortar business and was advised to network to be successful, one of the networking groups I joined was a local women's group called HeartLink, which was focused on connection and doing positive things for the community. While I knew one or two women initially, I did not know many of the women upon first joining. I quickly got to know them, however, because we met regularly, and with each subsequent meeting, I started sharing and relying more and more on them for feedback about different areas of my business.

Not only did this heartfelt and caring group of women provide helpful and positive feedback about branding and marketing questions that I had, but they also provided me with much-needed encouragement and support. I didn't feel so alone in this new and "somewhat scary" journey I had just embarked upon. These women became ambassadors for my business, and likely contributed to its quick uptake and success. Thus, I highly encourage you to get out in your community and network. You never know who you may meet who can connect with you on so many levels and connect you to other "helpers." Even just being able to talk about your goal and your mission can get you reignited and motivated, especially when you're hearing positive comments and excitement back from such a group!

## Other Support System Resources

You can also ask for help at your local library, health clinics, the YMCA/YWCA, or your church. Don't be shy. You never know who, or what, could come from making new connections. Most people want to help others. You just have to ask for it.[110]

## Deeper Support Systems

As I mentioned earlier, you will likely need different people for different types of support. You may need helpers who can offer deeper levels of support to you and more than lending a sympathetic ear. You will want helpers or pack members who encourage you to stretch yourself and push you to greater lengths, almost "locking arms" with you along your journey. These helpers are coaches, mentors, accountability partners, and mastermind groups. While they all sound like they may offer something similar, they are different and based on your needs. You may want to have them all in your pack and I'll explain why here.

## Coaches

A coach is someone who provides guidance to a client on achieving their goals and helping them to realize their full potential (sort of what I'm doing here in this book). It's a supportive role and one whose relationship is somewhat structured and often short-term. It is structured in the way of scheduling weekly calls, for instance, or check-ins with your coach to review your progress, discuss any hurdles you may have faced over the previous week, and get suggestions on how to personally improve or overcome the hurdles. When I decided that I wanted to walk on a fitness stage for the first time at age fifty, I knew I needed a coach. Not only for guidance in terms of a diet and workout plan, but to help get me ready for the big competition day without injuring myself or causing harm to my body. Each week I sent photos of myself to my coach and relayed any difficulties I had experienced during the previous week (i.e., low energy levels, headaches, missed workouts, injuries, etc.). From this information, my coach would then make small tweaks to my diet plan and/or give advice and needed encouragement to keep me on track. I know for a fact that I wouldn't have looked as lean and defined as I did on competition day without her coaching help.

There are coaches for different areas of life such as for finances, life, business, or fitness. Learn from their experiences, be a "sponge," and take what you need for as long as you need, from any or all of them, to help you grow personally and professionally in areas where you could use help. If you choose the right coach for you, you could not only save precious time in getting to your end goal successfully but become more of the person you aspire to be, because a coach stretches you and holds you accountable.

## Mentors

While a coach is more focused on performance, a mentor is more focused on skills. A mentor is one who shares their expertise to train a mentee in a new skill. For example, in a workplace setting, an older and more experienced worker may take a younger apprentice under his or her wing to transfer his/her level of skill to the less experienced worker. This is a more directive role where the mentor is teaching and the mentee is learning. This type of relationship is often more long-term compared to that of a coach.

Having a mentor is a great way to learn more expediently than on your own. You can also have more than one mentor, depending on what you want to learn. For instance, if you're wanting to grow your business, you may want to find someone who has had success already in growing a business and ask for their help. Offer to take them out for lunch and ask some well-thought-out questions to gain valuable feedback. If you enjoy this type of learning and can see how it can expedite your progress and learning curve (as it should), ask politely if you could set up quarterly lunch or dinner meetings to get their input on potential problems or hurdles you may come across in the future. (And of course, it goes without saying, that you will be footing the bill here for the lunch or dinner meetings in order to learn from their wisdom. It really is an easy and inexpensive form of education). Oftentimes, successful, more experienced people are honored to be asked for their advice and are often more than willing to help. It's a way for them to give back and makes them feel good.[111]

## Accountability Partners

An accountability partner is someone who holds you accountable. It can be like a mentor-mentee relationship because it involves two people, but it's often more of a peer-to-peer, two-way relationship where both parties are holding each other accountable to their goals, rather than a one-way in which an older person is training a younger person. In choosing the right accountability partner, ideally you want to try to find someone who has experienced what you're going through so he/she understands and can provide motivation when needed (and it will be needed).

Here are some tips to make accountability more successful:

1. Ensure you know each other well so that you're comfortable saying the tough things as well as providing encouragement. If you know the person well, it's easier to provide tough love and encouragement in ways that will resonate but not offend.

2. To get to know each other, you will need to schedule regular meetings, whether in-person, on the phone, or via video calls. Choose a schedule–weekly, biweekly, or monthly–that you both can commit to.

3. Establish rewards and consequences up front so it is clear to both parties what's on the line. Have fun with this. Choose rewards that will motivate you both and consequences that neither really wants so that you both don't fall off the wagon, but instead stay on track towards both of your goals.[112]

I have stated this before, but it bears repeating here . . . we often will work harder for others than ourselves, which probably explains why accountability works!

## PROBABILITY OF COMPLETING A GOAL

Here are some shocking statistics on the probability of completing a goal at various stages of commitment, gathered from a study done by the American Society of Training and Development:

| | | |
|---|---|---|
| 10 percent completion | ---→ | if you have an actual idea or goal |
| 25 percent completion | ---→ | if you consciously decide to do it |
| 40 percent completion | ---→ | if you decide when to do it |
| 50 percent completion | ---→ | if you plan to do it |
| 65 percent completion | ---→ | if you commit to someone that you will do it |
| 95 percent completion | ---→ | if you have a specific appoinment(s) with an accountability person that you have committed to[113] |

Not only do these statistics reveal the importance of having an accountability partner, but they also highlight the importance of the Dig and Leap sections in this book in getting you to decide on a goal, making a Leap Plan with dates included, and selecting helpers along the way.

## When It's Really Important to Have an Accountability Partner

One thing to point out, too, is that having an accountability partner is especially important when you're stretching your comfort zone (which you should be doing anyway as I outlined in the Stretch like a Dachshund chapter) and/or are learning new things. If we're just relying on ourselves in these situations, it's easy to give up on ourselves when we find things getting difficult or uncomfortable, but less so when someone else is involved whom we don't want to disappoint.

## Accountability Partner as a Coach

An accountability partner can be a peer but can also be a coach to whom you report to on a regular basis for motivation along the way.[114] In the case of my fitness journey, my fitness coach was also my accountability partner. She provided me needed advice and encouragement weekly, and I, in turn, did not want to disappoint her. But truthfully, I had also wanted to prove to myself that I could do it, so I was accountable to myself. I was also accountable to my friends (and their friends) on social media because I had posted and chronicled my fitness journey online. I knew if others were watching my progress, I would work harder . . . and I did. It was like I had hundreds of "invisible" accountability partners who were subconsciously pushing me–some of whom were rooting for me, but some who were perhaps secretly hoping I would fail. I didn't give them that satisfaction. I knew that by posting my goal and journey online, that 'act of vulnerability' would ultimately help me succeed, and it did.

The surprise factor for me, however, in posting my journey online was just how many hundreds, if not thousands, of people across North America, both men and women, got inspired to get healthier and fitter. This was a thrill for me (like an inherent surprise reward) that I hadn't anticipated happening.

## Mastermind Groups

If having a coach, mentor, or accountability partner can bring you advice, feedback, knowledge, resources, connections, and motivation, imagine what a group of people could do for you? This is the power and premise behind mastermind groups.

Napoleon Hill, author of the classic book *Think and Grow Rich*, first wrote about the concept of mastermind alliances and the mastermind principle in the late 1920s in his first book, *The Law of Success*. He wrote that when two people get together to work on a problem, a third mind is created: the master mind, a separate and somewhat energetic force.

Mastermind groups started developing that were formed as friendly alliances of people from which to learn and grow from one another. By borrowing and using the knowledge, experiences, and capital (in some cases) of other people in the group, there would be an increased chance of being able to carry out your plans. The mastermind group could also form a collective mission for the group, as well as help one another with each of their own goals. So, there are really two inherent purposes for such groups. As Hill stated in his book and personal development program, one could accomplish more in one year with a mastermind group than in an entire lifetime without one! [115]

Some examples of famous mastermind groups are:

1. The Vagabonds, which was made up of Henry Ford, automobile mogul, Thomas Edison, inventor of the lightbulb, US President Warren G Harding, and Harvey Firestone, founder of The Firestone Tire and Rubber Company, who from 1915-1924 took caravan road trips every summer to hang out together at campsites in the Eastern US to explore nature, seek inspiration from one another, and discuss scientific and business ventures, all with the primary goal of improving themselves and solving problems for their companies and others.

2. Famous English writers and poets C. S. Lewis, J. R. R. Tolkien, Charles Williams, and Owen Barfield met regularly to discuss their respective works in progress and produced such famous works as *The Chronicles of Narnia* and *The Lord of the Rings*.[116]

Hallmarks of a good mastermind group are that it:

- Is friendly and harmonious

- Is growth oriented

- Is supportive, with members willing to participate and share

- Has a group size ideally of five to ten people from either a similar or different business or background

- Involves a commitment of time and energy[117]

A mastermind group can bring many benefits, such as:

- Expanding your view of the world

- Providing advice and feedback

- Helping you solve problems

- Challenging you to set and accomplish your goals

- Helping to generate ideas

- Encouraging collaboration amongst the group members

- Hearing others share their successes and challenges

- Providing personal and professional development opportunities

- Creating a safe, supportive, and learning environment for all [118]

With so many positive benefits to being involved in a mastermind group for both personal growth and support, the time and energy commitment that would be necessary would be well worth it. And imagine if there was a collective mission for the group that each of the members contributed to? The end results could be outstanding!

I hope I have shown you here, both in nature and in the human world, the importance of having a support system or pack. We can do, and become, so much more when we're surrounded by others who want to help us succeed. We can often accomplish so much more when we're part of a team. So be sure your helpers in your Leap Plan are either mentors, coaches, and/or accountability partners. Ideally, have one or more of each, and for sure, consider finding or starting your own mastermind group. The results could blow your mind and have you completing your goal(s) in a shorter time than doing it alone.

Plan your pack carefully and then run together to your goal line!

## FETCH STEPS

1. If you haven't yet identified a coach or accountability partner as part of your helpers, then take the time to seek, interview, and find one this week. Be sure you're comfortable with them and can trust them. Share your goal with them and commit to meeting regularly. Add your scheduled meeting times in your Leap Plan and/or your calendar.

2. Find or start a mastermind group. Here are two guides to aid you in starting your own mastermind group:

   a. "How to Design a 90-Day Mastermind Group," a blog entry written by Jack Canfield on The Success Alliance website.[119]

   b. "Six Proven Steps to Forming a Mastermind Group" by Eventual Millionaire[120]

3. Read or listen to one or both of the books authored by Napoleon Hill, **Think and Grow Rich** or **The Laws of Success**.

# 15

# Be Alpha

**Now that you have formed your Pack,** your support system, I want to encourage you to be the alpha of your pack. I'm not talking about being domineering and barking orders to the members of your pack here, but more to encourage you to think like, and be, a leader.

By being the leader of your pack, you are in control of your life and your destiny. Basically, you're in charge of YOU. Live life according to your rules and wishes, not someone else's. While you may not have all the answers, call upon your mentors and coaches to save time and learn from them to not waste time "reinventing the wheel." Stay focused on your goal, follow your plan, and leverage the Helpers in your Pack, when and where necessary, to help you get to your goal.

## Lead Yourself

With your Leap Plan as your roadmap, your Pack for support, and with focus and determination, you can do what you have set out to do. Learn to lead yourself. Stay focused on your goals. Work towards your rewards and your end goal. Keep the vision of what you see for yourself as your driving force. You can do this. If not you, then who? No one else can do this for you. You need to do this for **yourself**! Get help when you need it, but don't give up. Think like Winston Churchill and "never give up." Don't give up on yourself. When you lead yourself and accomplish your goal, then you can go one step further and lead others to do the same.

## A Leader of Others

As your "pseudo-coach" in this book, I want to stretch you further beyond leading yourself, to consider becoming a leader of others. Why? Because we need more leaders. We need leaders who are doing good for society and the world; leaders who are using, shining, and sharing their unique gifts for all to enjoy and benefit from. This is how we all grow both personally and as a society.

The point of this chapter is to get you thinking like an alpha, a leader, and taking the learning process from this book to go beyond yourself and your goals to those of others. Learn to lead others. While it may not be for everyone, I am encouraging you to consider stepping up to the plate to become a leader. Whether in your business, at the workplace, or in your community, having a hand in leading and developing others to see their potential will surely bring you greater rewards than you can imagine.

Before I delve into the qualities and benefits of being a leader, I want to address why I want you to think like an alpha and to also dispel some of the myths around alpha males and alpha females.

## Why Alpha?

Alpha is the first letter of the Greek alphabet and is a synonym for dominant.[121] The term "alpha male" first originated from the field of animal behavior and refers to the highest-ranked individual who rules the social group. Being alpha, whether male or female, means being the undisputed leader of the pack. We see this hierarchy in packs of wolves, baboons, lions, and humans.[122]

Interestingly, in packs of wolves, it used to be thought that the alpha male and female of the pack exerted dominance and aggressive measures to "show who was boss," but that was only observed in wolves in captivity. Over the past decade, studies of wolves in their natural habitat have shown that alpha males and alpha females are simply the breeding animals. They are the "parents" of the pack, and dominance interactions are rare. A typical wolf pack, then, should be viewed more as a family with adult parents guiding the group activities, such as capturing prey and protecting and caring for the young. So, as an alpha, in essence, you are taking care of your pack.

In the wild, all young wolves are potential breeders, and when they mature and breed, they automatically become alphas, break off from the original pack, and start their own pack (not unlike our children growing up and leaving the nest to start their own family). Certain animals, like dogs and wolves, fare better in a pack because they look for a leader to lead them. In both a wolf and dog pack, there is always an alpha dog who is seen as the top of the pack. The other pups or wolves look to the alpha dog for leadership, structure, and protection.[123]

## Dogs as Pack Animals

Dogs are social pack animals. They like to belong to a pack and will seek whichever pack is nearby. When a new dog is brought into a family home, for instance, whether or not there are other animals present, the entire family becomes part of the pack.[124]

Just like animals want to be led, so do many humans. We like leaders and we follow them. Look at famous people like actors, musicians, fashionistas, and innovators and how many followers they have on social media. Some of them have tens of millions of followers, and some of those have no skills or special gifts!

You *can* be a leader. You have knowledge, skills, a gift, mentors, and a support system around you. You've built up your confidence by creating a Fetch mindset where you're willing to put in the necessary work with focus and dogged determination. You have a goal and have created a plan to get there. Now you just have to do it.

## The Leader Process

All I am asking is to follow this similar Dig. Leap. Play. process to become a leader. Have a vision, set a goal, create a plan, consider the helpers and hurdles that you may face and need, stay focused on the goal, share it with your group, and encourage your team with rewards and positive reinforcements. It's basically the steps that you are learning in this book, except instead of just focusing on your goal, you are applying it to a shared goal for the betterment of the group. The process is essentially the same. As their leader, too, if you can help the members find their gift, special skills, or knowledge and leverage these amongst the team to more easily accomplish whatever goals are in place, imagine the power, ease, and results that can happen here. It could be monumental! Can you imagine how good you would feel knowing you had a part in this? This is why I am encouraging you to be a leader.

Here are some more benefits of being a leader:

1. You can exponentially bring more impact to, and positively influence, others.
2. You will grow personally in terms of your capabilities, confidence, and competence.
3. It's fulfilling to help and serve others.
4. Together, with a team, you can create positive change.
5. You can build hope, trust, stability, and compassion in others.
6. You can leave a legacy by leading and teaching your ways . . . or your gift.

Everyone has the potential to be a leader. But it does take time and hard work to develop good leadership skills. Through regular disciplined learning, a desire to improve, a positive attitude, and a concern for others, you can become a leader. Learn by reading or listening to leadership books and podcasts, attending leadership seminars, and . . . by leading others. You may trip and fall at first but keep trying. Be creative; be resourceful. Ask for help from your coaches, mentors, or other leaders. Apply the CANI principle to leadership–keep learning and growing.

## Lead with Your Strengths

If you focus on improving what you're already good at, such as your gift or other natural talents, then you can lead with your strengths. For example, my gift is to connect and communicate, whether in writing or speaking. I choose to lead through my words; leading and inspiring others to step up to their full potential through my books or my talks. If my perspective, knowledge, and experience can help others take action to finally live out their dream lives, or to at least do that thing they have always wanted to do because of me, then I will feel fulfilled. I will feel that I have lived my purpose here on Earth, because that is what is in my heart to do.[125]

I challenge you to be strong in who you are, or who you want to become. You are unique. Stand in your light and shine your gift(s). By using and sharing your gift, you may find that people will naturally follow you.

## Lead with Empathy

If you want to help others, as I do, then lead. Lead with empathy. Show you care about them first. Think of others before yourself. Be here to serve others. This is what will bring you true fulfillment and joy, especially if you are using your gift to do so. Find ways in which you can use your gift to help others. Get creative. Brainstorm with others, your mentors, your peers. You may be surprised by what ideas you collectively come up with. Remember, there is strength in numbers.

By sharing your gift, you are essentially leading with your heart. You are putting your authentic self out for the world to see, which is good. We not only need more leaders in the world, but we also need more authentic leaders. Leaders who are empathetic, enthusiastic, trustworthy, and inclusive. Leaders who have integrity, vision, are good listeners, and have good communication skills (here's where those Toastmasters classes can come in handy).

Leadership is about people. It's about inspiring others to use their talents (i.e., gifts) and move them in a direction to achieve certain outcomes or toward a common goal.

*"The people's capacity to achieve is determined by their leader's ability to empower. Only empowered people can reach their full potential."*
– JOHN MAXWELL

And, as you've already learned earlier in this chapter, we can empower others by recognizing their strengths or gifts and using positive reinforcement. An inspired, motivated individual can do great things. Imagine what can get accomplished when you inspire a group of people.

## Genuine Leadership

Just like the alpha in a pack of wolves is not about being dominant or aggressive, leadership is not about controlling or intimidating others. Genuine leadership is about using your gift or influence to inspire and motivate others to do great things.

As an alpha, you have the courage, determination, and energy to drive change, people, and results, whether in your business or at the workplace. Yes, you will have to take on more responsibility as a leader, but with that role comes a thrill of knowing that you are growing into the person you were meant to be, as well as making an impact. Trust your (animal) instincts and learn to make quick decisions (strong qualities of a good leader). Lean on your coaches or mentors for help in leading others and remember to put the goals of your team or followers ahead of your own. You'll earn their respect and likely gain more followers by doing so. Be a good listener but stay focused on whatever the goal is for the group and work together to achieve it.

## Develop Other Leaders

Leverage your own past experiences and successes as well as the skills and experience of the group that you are leading. Be inclusive and empower others to step up to the plate. They will grow in the process, as well. Become a leader that develops other leaders–the results will multiply many-fold for both you and them. By enlarging others, you become larger in both your character, abilities and connections and thus, can potentially make an even bigger impact.

## Requirements of a Leader

If being a leader excites you, then I would encourage you to read some of the many books on leadership. Author John C. Maxwell is the guru on leadership. One of his books, which I have read more than once, is *The 21 Irrefutable Laws of Leadership*. It's an older book now, but it's a classic, especially on leadership. In it, he outlines the seven requirements for being a leader. They are character, building relationships, knowledge, trusting your intuition, experience, past successes, and ability.

In terms of character, a leader should be a good person on the inside. He/she should have integrity and be respected and credible. A leader knows that in order to lead others, you have to get them to first trust you. In order to gain trust, you have to spend time to build solid relationships with the right people. They have to know you as a person first and ensure you are leading for the right reasons.

A leader needs to be knowledgeable. They need to have the right and sufficient information in order to gain the trust of the group, as well as a vision for the group to buy into. The vision is what keeps the people going. A leader needs to be able to trust his/her intuition, whether it's something that he/she is born with or develops, in order to be able to assess certain situations when leading and coming up against hurdles or roadblocks.

Being able to draw on past experiences and successes shows competence in a leader and builds confidence in the group toward the leader. In essence, the leader has been through the "trenches" already and knows what to expect so he/she can best guide others in the process. And finally, a leader should be capable of leading–being able to provide direction toward a goal as well as a sense of purpose in order to make positive change.[126]

To assess if people are thinking differently now about leaders, compared to how Maxwell viewed them twenty-five years ago, I ran a survey on social media. I asked participants for their opinion about leadership, namely what qualities they admired about leaders as well as what they wanted to see in leaders. Here were their (and my) top ten answers:

1. Action-oriented
2. Articulate
3. Confident, competent
4. Empathetic, compassionate
5. Understanding
6. Encouraging
7. Trustworthy
8. Courageous
9. Respectful
10. Visionary

I like this list because it more aptly describes how a leader should behave towards others and is more inclusive and considerate of others. While we all like someone who is bold and courageous, we do like a leader who is kind and caring. We can get behind someone who understands us and is making decisions and taking the actions with everyone's needs in mind. If you agree and have some or all of these qualities, then definitely consider becoming a leader. The world needs you.

I had also asked in the survey where they felt we needed more leaders. The general consensus was that we need more leaders in these following areas:

1. Schools
2. Community
3. Politics
4. Workplace

So, take your pick as to where you would like to exert your leadership skills. If you are passionate about helping out children or teachers, have ideas about the curriculum, or simply want to lead a group of parents at your children's school, then consider volunteering or working within the school system. If you see an opportunity for yourself where you can exercise your gift for your community, then I encourage you to reach out to the many associations, groups, and/or churches present in your local community. I am confident that they would love your help and vision. If you enjoy politics and want to make a change that could impact a larger group of people or be the voice for a specific group of people, then get involved in your local political group. And of course, I am sure that your workplace has opportunities as well to get involved and lead. Show initiative and reach out to your superiors. Share your ideas and vision and suggest where you can lead others. Stretch yourself and speak up. You will learn and grow personally in the process and be proud of yourself for the impact that you can make on others. You may surprise yourself when you see what you are capable of!

In summary, leaders inspire, align, and motivate. If you are inspired to be a leader, then I celebrate you in advance. Step up to the plate, take charge, and make positive change by leading others! What you get in return by doing so will be incredibly fulfilling for you. While this, in my opinion, is the ultimate in fulfillment, learn in the next chapter how we can reward ourselves in other ways.

## FETCH STEPS

1. Take this time to revisit who is in your own Pack. Are they letting **you** lead or are they taking charge and pushing their goals and agenda on you? Are they supporting you as you had hoped? If not, perhaps you need to find others who could, or simply do not reach out to them. You do need to be somewhat selfish here and ensure that those in your Pack are supportive of your goals and you.

2. Read some books on leadership, like John Maxwell's *The 21 Irrefutable Laws of Leadership,* and/or take some leadership courses online or at a local library or college. The benefit of taking a course, especially an in-person course, is that you also get to meet other leaders who you may want to reach out to in the future if needed to brainstorm or to collaborate. Perhaps even start a Leadership Mastermind group in your community!

3. Start small. Is there a leadership role you can ask to be a mentor for at your work? Can you volunteer for a leadership role in your community or church? Get your feet wet first and then take on bigger leadership roles as you build your confidence and leadership skills.

# PLAY

# 16

# Give Yourself a Treat

**When I say, "Give yourself a treat,"** I am not referring to treats here as just food to simply feed you (but they could be), but more to refer to using treats synonymously with rewards. Humans and animals will do just about anything for a reward. We all love to be recognized for doing something good. We enjoy winning; we enjoy earning a reward. It makes us feel good. It's a treat to accomplish something, i.e., achieve a goal and reap the reward for doing so.

In creating your Leap Plan, you get to decide which rewards will drive you. I want you to use mini-rewards along the journey to your Stretch goal, as well as a major reward for when you reach your end goal. This chapter will focus on why treats or rewards are used for both humans and animals, and how we can use them to our advantage.

For some people, just the act of attaining a goal is a reward in and of itself, and it is. However, for many of us, we like to reward ourselves in a more ceremonious way–by celebrating the win and enjoying the reward that *we choose* and like. Rewards are also used as icing on the cake. We want to attain our goal (the cake), but the reward is like icing. It's even better, maybe even more enjoyable, than the goal itself. The rewards serve to drive us and keep us motivated to keep going.

This chapter will show how rewards are used to help animals learn and direct behaviors, just like they can with humans–both adults and children. We'll learn the types of rewards, the timing of them, and when to use which ones where. We'll also get a better understanding of why rewards work so well and why you want to ensure you incorporate lots of rewards in your Leap Plan.

## Positive Rewards for Animals

In the veterinary behavioral world, the shift over the past few decades luckily has been to promoting reward-based training rather than the aversive type of training from the past. Pets learn best and more quickly when motivated and rewarded by something they like, something positive.

Food is the easiest for us to use and is a great motivator for training animals, especially when they are food-driven and hungry. Reward-based training involves rewarding desired behaviors and removing rewards for unwanted behaviors. For example, feeding treats within one to two seconds when a dog or cat sits rewards the animal for calm, sitting behavior (the desired behavior for the human, that is). Animals learn to perform those behaviors that are reinforced (i.e., treated). The goal for pet owners is to ensure we are rewarding the right behavior at the right time, otherwise we could send confusing messages to the pet.

As I mentioned earlier, when working in the clinic, I carried a treat bag filled with tasty (I assume they were), bite-sized, moist liver or beef treats so that when a dog entered the reception area, I would give him or her a yummy morsel–just for entering the clinic. This food offering, in most cases, instantly decreased the pet's fear, distracted him/her from other pet smells and sights in the clinic, even momentarily, and made walking into the clinic for the dog (and probably the pet owner, too) a more positive experience from the beginning. I would then use the treats as a reward for good behavior like sitting calmly or letting me palpate its abdomen (the pet's, not the human's) in the examination room. Owners would remark that on future visits, their dog was excited to come in. Even the initially anxious ones were less fearful and more curious on subsequent visits. They began to associate the clinic with receiving treats, which made the clinic visit a more pleasant experience.

Research studies have indicated that positive experiences with the veterinarian, even during a simple meet and greet, can actually protect pets from being adversely affected by unpleasant procedures experienced at later dates, which is beneficial for keeping the stress load down for pets and the pet owner, as well as reducing hazards for the clinic staff who are involved in handing the pet.[127] Referred to as "latent inhibition," this means that for the pet, the first appointment at the clinic will be remembered with more significance than subsequent appointments. In other words, positive first visits will overshadow the "bad" ones in which they may receive a vaccine, have blood taken, or have their anal glands expressed, all of which can cause some slight discomfort, even if momentarily.[128] So, the moral of the story here is to bring your puppy and kitten into the veterinary clinic at an early age, say around twelve to sixteen weeks (which is also within the socialization period for pets and the best time to socialize them to many different sights, sounds, and experiences), even for just pats and treats, to socialize them positively to the clinic environment and staff early on! The full socialization period for cats is three to nine weeks, while for dogs it ranges from three to twelve weeks. Continued periodic socialization up to age six to eight months is recommended for dogs to prevent regression.

## The Use of Force-Based Training

The use of force-based training or punishment involving corrections such as choke chains, pinch collars, forcing puppies onto their sides or backs (submissive positions), or spray bottles to punish unwanted behaviors often results in puppies shutting down and not learning. The use of such aversive techniques can also escalate a dominance problem, have major negative effects down the road, and be potentially very dangerous for the human trainer or pet owner.

From science-based training methods over the recent years, we now know that rewarding desired behaviors with food and other positive rewards works much better and are longer lasting for our pets.[129]

Similarly with the human species.

## Positive Reinforcement for Humans

Positive reinforcements go far with people, too. Positive reinforcements such as gifts, incentives, recognition, and even compliments go a long way to make people feel appreciated, and in many situations, will make a person repeat that rewarded activity or behavior. For instance, if you were to compliment someone on their shirt or tie, chances are good that he or she will reach for that shirt or tie again soon, having been influenced by your kind words and remembering how good your compliment made them feel.

Positive reinforcement also works for children. Rewarding behaviors that parents and teachers want to see a child do increases the likelihood of that behavior being repeated again and more often by the child. Rewards like praise, clapping and cheering, or giving a hug or a pat on the back are just some forms of positive reinforcement that children enjoy seeing, hearing, and feeling. Other forms can be more tangible, like a toy, special privileges, or taking them to the playground for doing things like cleaning their room. Kids who receive positive reinforcement are motivated to continue working hard and doing even better the next time—all good habits to instill.[130]

At the workplace, adults will often work harder for recognition than money. Business and employee morale studies have shown that employees want to be recognized. They often want it over more money, more interesting work, job security, or good working conditions.

Research done in 2012 showed that 56 percent of those who had received recognition considered themselves highly engaged compared to the 33 percent who didn't get any credit for their work.[131]

Charles Schwab, a steel magnate and one of the first people in American business to earn a million dollars a year back in the early 1920s, said that he earned this salary not because he knew more than anyone else about steel, but because of his gift in knowing how to deal with people. He learned quickly that the best way to encourage enthusiasm and bring out the best in his people was through appreciation and encouragement. This leadership philosophy still holds true today, even one hundred years later.[132]

## Types of Rewards:

So, what types of rewards work to give positive reinforcement?

The four main types of rewards are:

- Food

- Praise

- Affection

- Materialistic–i.e., gifts and incentives

All four types of rewards can work for both animals and humans, except incentives. Incentives like money or trips obviously will only work for humans (although some dogs get pretty excited about trips to the park), however, the other main types can work for animals as much as for people.

## Food as a Reward

Food, being a necessity for survival for both species, is a strong reward for both humans and animals, and especially if it's tasty. In my own situation, I worked hard on my fitness and diet plan every week for that yummy piece of chocolate cake at the end of the week during competition season. Dogs and cats, especially if food-motivated and slightly hungry, will work hard to do what you ask to get that morsel of food. Often, parents will reward a child with a dessert or a trip to their favorite restaurant if they do a particular activity like cleaning the house or achieving a certain grade on their report card. Food can work wonders to drive the correct behaviors and habits in both animals and people. Are you using food as one of your mini-rewards in your Leap Plan?

## Praise as a Reward

Throughout our lives, we strive for verbal praise from our parents, teachers, bosses, and colleagues. A few kind words can go far in making us feel good about ourselves (and they're less fattening than chocolate cake!). Animals will also equate praise such as a pat on the head or the words "good boy" with positive reinforcement, especially if it was paired in the beginning with food. Eventually, they learn that verbal praise or a pat on the head is as good as the food (more or less). Verbal praise from others, especially when unexpected, is even more special. Praise like, "Wow, you look great in that dress!" is extra rewarding because it's unexpected (and greatly appreciated).

Praise in the form of expressing our appreciation for someone else is incredibly important and should not be undervalued here. These days, we all get so busy doing our own things and working away that we can easily forget to show our appreciation for others. We fall into the trap of taking others for granted–our parents, our kids, our spouse, our staff. It is important to be conscious of showing appreciation for others on a regular basis. The good that it brings others will surely bring rewards back to you in the form of smiles, happiness, a good family and social life, and much love. Saying that you appreciate someone as well as showing it in small acts of kindness, like making your spouse breakfast or taking them out for lunch, or buying them flowers for no special occasion, will certainly go a long way to making them feel appreciated, wanted, and loved. Appreciation in a marriage is as necessary as expressing love. Many families break down because of the lack of appreciation shown by their loved ones. By not acknowledging someone or their worth, the person can feel devalued, disappointed, and less motivated to work on their marriage, for instance, or their work. Get in the habit of giving appreciation often and where necessary, especially to those close to you.

## Affection as a Reward

Affection by others, as a form of praise, fills our need for love and belonging, which is one of Maslow's hierarchy of needs. As you will soon learn, when we receive affection, usually in the form of a hug by a loved one or a kiss by a spouse, oxytocin is released which makes us feel good. Children, too, just want to be loved, and showing affection as a form of positive reinforcement will not only encourage a child to repeat the desired behavior

but will also make them feel wanted and loved. If affection or attention is what you desire at the end of your goal, then make sure you are visualizing yourself daily feeling those emotions that you want to feel when you reach your end goal. Put it out into the universe to attract it to you.

## Materialistic Rewards

External rewards like gifts, toys, and things that have to be bought (in most cases) can also serve as a form of positive reinforcement and motivation. A gift can drive one to reach a goal, especially if it is highly desired. An incentive trip to somewhere highly desired or a special outfit may be all that you need to motivate yourself to complete your goal. Ensure you have set a special major reward for yourself when you reach your end goal on your Leap Plan. Having smaller materialistic goals to reward yourself along the way to your end goal, like massages, pedicures, or a new pair of running shoes, for instance, can keep you motivated and on track to your end goal. It also keeps it fun and fresh!

The key here, of course, is to only reward yourself when the goal has been achieved. In other words, you need to earn the reward. Do not cheat and give yourself a treat if the goal has not been attained. This dilutes the value of the reward. Be sure your Leap Plan states what the daily or weekly goals are for you to try to attain and what the mini-rewards are once they are attained. Reward yourself only if the goal has been achieved for the week. An even better idea may be to have someone else provide the reward, like your coach or accountability partner, if, and only if, you have attained your goal (and it's been verified).

## Give and Receive Compliments

Knowing how much a compliment can brighten someone's day, be freer with giving out compliments. Give them out like Halloween candy and give them often. Give them out to colleagues, staff, family, your children, and even strangers. When you receive great service or a tasty meal at a restaurant, compliment the server and the cook. Recognizing someone for a job well done goes a long way to building that person's self-esteem. People generally want to hear that they are doing a good job and that their efforts matter. Whether it's a young teenage girl dancing a solo at the studio, a son building a go-cart in the garage, or a grandmother cooking a pie for Thanksgiving, they all like to hear that their hard work matters and is appreciated. A compliment and recognition of their efforts will likely encourage them to do it again and enjoy doing it even more than the first time!

Get in the habit of looking for things to compliment others on, especially when meeting people for the first time, as well as to your family on a regular, if not daily, basis.

When you first meet someone and want to set a good first impression, compliment him or her. Perhaps you like their hair, jacket, or a piece of jewelry that they are wearing. Compliment them on it. And of course, smile at them, shake their hand, and certainly remember and use their name. That person will more than likely instantly warm up to you and will likely remember you, as well.

Smile at and compliment your kids in the morning before they leave for school. Saying something like, "I like your hair, honey," or "You're looking sharp today!" will surely make them feel good about themselves and may set a positive tone for the day. Same goes for your spouse. Telling him how attractive he looks wearing his new shirt will probably have him walking taller all day at work.

And when *you* receive a compliment, don't shrug it off like it's nothing. Accept it graciously, sit with it, and enjoy it. Take the compliment as recognition for a job well done. Feel good about yourself and see how that compliment makes you feel. Please do not downplay any compliment sent your way. It short-changes you and the person giving it.

## Be Sincere

Any compliment that you give, of course, should be well founded and sincere. Do not make things up and fall into the trap of just saying what the other person is wanting to hear. This is flattery and it can often be shallow and insincere. While some people may enjoy flattery because it feeds their ego, most discerning people do not, as they can sense that it is not from the heart, like true appreciation is.

All humans hunger for appreciation. It is like food for our souls and makes us feel like we matter. Thus, make a habit of sprinkling these little words of encouragement and gratitude in your daily life for others to enjoy and be lifted by. You may find that you are lifted too in the process!

## Timing of Rewards

When training a dog or cat to learn a complex trick, we give rewards as they get close to what we want them to do, even when they haven't reached the final step. This process, as I have alluded to earlier in the book, is called shaping or successive approximation (remember the example of the dog learning how to do a somersault?). Essentially, we are rewarding the "baby steps" to the final goal to help guide our pets in understanding what we want, and to help us stay on track toward our goal, as well.

Similarly, we do not need to wait for perfection or the final goal to reward our kids, family, staff, or ourselves before accepting a reward. Rewarding in small ways encourages and motivates them (and you) to continue to do better and to stay on track. If you have just reached your goal of losing ten pounds, for instance, on the path to losing fifty pounds, then go ahead and buy that pair of jeans you've been wanting or treat yourself to that much-needed massage. You will feel a sense of accomplishment already and be given a sample of what reaching your end goal could feel like. Using mini-rewards to reward yourself along the path to your end goal will help keep you motivated. These little wins keep us positive and encouraged to keep going.

This same concept holds true for financial success. Many experts say to "pay yourself first" on the path to financial success. By getting in the habit of paying yourself first, and saving or investing for something for yourself, your savings and net worth will grow, and the habit will create more confidence, discipline, and creativity for you–all good things.

## Celebrate Your Successes

It is important to celebrate your successes, no matter how big or small. Just like with your strengths and your gift, your successes are easy to take for granted. If you don't treat yourself occasionally, it can begin to feel like all work and no play, and where's the motivation in that, right? (More on the importance of play in the next chapter). Unless, of course, your work does feel like play to you, and then you are one of the lucky ones!

You need to feel good about yourself. Acknowledge your strengths, skills, and achievements. Even if others don't notice it, you should. It will serve as fuel to continue doing more and striving for even bigger goals than what you laid out in your Leap Plan. The chapter entitled Chill Like Your Cat will get you thinking about the next step once you've completed your end goal in your Leap Plan, so stay tuned. You need to love yourself. Focus on your good qualities, your gift, the special things you do, and the positive things you bring to the table and for the world to enjoy. Believe in yourself and your gift. You *can* make an impact and you *do matter*.

## Why Do Rewards Work?

So why the emphasis on rewards and what's the big deal? What is it about rewards that makes us work harder?

Simply, animals and humans like to do things that make them feel good. What makes us feel good about receiving a reward, whether a material good or verbal praise? Natural chemicals in our brains such as dopamine and oxytocin are released when we receive a reward. These chemicals or messengers hit our brains, and those of our animal friends, triggering feel-good emotions. Yes, it has been shown now that animals also feel positive emotions.

## Effects of Dopamine

When animals or humans receive a reward, whether expected or a surprise, dopamine neurons are activated that gives us a feeling of pleasure. We repeat the behavior that activated the dopamine neurons because we want those pleasurable feelings again.

## The Brain's Reward System

There's actually a reward system in our brain that drives our behavior towards pleasurable stimuli such as food, sex, and alcohol, and away from painful ones like conflict or homework, etc. This reward system, a group of structures at the core of our brain, processes emotions and sends them to our prefrontal cortex, where we then make decisions about whether or not to start or stop an action or behavior. Dopamine, which makes us feel happy, is also known as the "go get it" neurochemical. It acts as a signal to make us move; moving us towards rewards or behaviors that we need for survival (like food, reproduction, and energy conservation), so we repeat these behaviors because it is for our own (species') good.[133]

### BASIC NEEDS FOR A SPECIES TO SURVIVE

1. **Food**

2. **Reproduction**

3. **Energy Conservation**[134]

Dopamine also helps with memory retention, which explains why animals and humans remember what behaviors elicited the positive emotions and are motivated to repeat them.[135]

## Effects of Oxytocin

Oxytocin is a natural hormone that can be released through skin stimulation such as touch. When we get a hug from a parent or a spouse, it feels good because oxytocin is released and promotes positive feelings. In fact, oxytocin is often referred to as the "cuddle chemical."

Interestingly, this "feel good" effect is felt by both partners in the exchange and can happen between a dog and an owner. When an animal is stroked and petted, oxytocin levels increase in both the pet and owner.[136]

So again, we repeat behaviors that will bring us these positive emotions. And who doesn't love a hug, right?

## Keys to Rewards

The four main keys about rewards and how to use them for directing behavior is to:

1. Be consistent
2. Reward close to the desired behavior
3. Ignore bad behavior
4. Use a desired and positive reward

Being consistent with using rewards to direct behaviors is especially important when training animals to do new behaviors or tricks. The rewards are used to shape behaviors through a series of actions to the final desired behavior as well as to solidify in the animal's brain what behavior is desired. This is why consistency is so important. It's through repetition, initially, that the pet learns what behavior is being desired and rewarded. I would also recommend, however, that you, too, be consistent in using mini-rewards along your journey to your end goal. When the journey is shorter, i.e., in reaching a mini-goal, it keeps it fun and rewards you sooner rather than waiting until you reach the end goal, which could be months down the road. If you are like me, I am impatient and want a treat sooner rather than later, no matter how small! A treat is a treat.

By rewarding yourself (or your pet) closer to the accomplishment of the desired behavior, then the behavior that got the reward will likely get repeated. For example, if you reward yourself with a dip and swim in the pool after a five-kilometer run in the summer heat, and it makes you feel amazing afterwards, then chances are good that you will do the same again the next time. Giving a reward close to the desired behavior teaches not only your pet what behavior to repeat, but also teaches you to rejoice in the joy of accomplishing your goals (even the baby steps along the way).

If you happen to "slip up" on your journey to your goal by doing something "bad" like eating a cookie while on your diet, or not going to the gym like your Leap Plan had stated for you to do, don't beat yourself up about it. Get back on track the next day. Avoid berating yourself or talking negative talk to yourself for not sticking to your plan. This will do you no good. Such negative talk can be demotivating and discouraging. We all have slip-ups; we are human. Simply recognize it as such, perhaps talk to your coach about it, and focus on the next steps. Focus on what you need and want to do and then reward yourself for doing the "good deeds." The same goes for our pets. We reward the good behaviors and ignore the bad ones. They eventually learn that the good, desired behaviors get the rewards. Both humans and our pets work hard for those rewards, especially when they are rewards that we want.

Now that you have learned why rewards and positive reinforcement work, the types of rewards that are available, and how best to use them, be sure to go back and change some of your rewards in your Leap Plan, if need be. Ensure the rewards that you choose in your Leap Plan, for both your mini- and major-rewards, are truly what you want. They need to be motivating for you to do the necessary, sometimes hard work to get them. They should also be positive and constructive, not something that is detrimental to our health like drugs or alcohol.

The benefits of rewards are many. By rewarding yourself along your journey to your end goal, through mini-rewards at different benchmarks in your Leap Plan, as well as a major reward at the end, you are helping to improve your self-confidence, giving you a feeling of accomplishment, keeping you focused on the task at hand, reducing procrastination, and improving your motivation and productivity.

## Benefits of Rewarding Yourself

1. Improve self-confidence

2. Gain a feeling of accomplishment and have less procrastination

3. Remain focused on the task at hand

4. Increase productivity and motivation[137]

Please don't ignore this important part of your Leap Plan. Reward yourself throughout and often. (There is no danger in using too many rewards, by the way.) Use it to your advantage and get ready to be surprised when others give you compliments, attention, and appreciation. Take it and keep going. You are well on your way to reaching your big goal!

Learning to reward yourself is learning to appreciate *you* and enjoying the little things in life. One other thing that I want to remind you about is to have fun. The next chapter will highlight the benefits of playing more, which I'm confident you will enjoy, just like our pets do!

### FETCH STEPS

1. Give someone a compliment today. Be sure it is sincere. Notice both their reaction and how giving a compliment makes you feel.

2. Go back and review the rewards you wrote in your Leap Plan. Did you include as many as you should? Are they motivating enough? Revise them to drive you. It's *your* Leap Plan—make it work for *you!*

# 17
## Play More

Why should I mention incorporating more play into our lives? Because it has so many benefits, it's fun, and I believe we need more of it in our lives. Just like with rewards, play brings joy and happiness for us and . . . it's pleasurable!

## All Things Play

Play can be an activity or a state of mind. It can be a means of self-relaxation or a reason to bring people together. Play is good for the mind as well as for the body. It stimulates the creative juices, brings joy and laughter, and improves one's mood as well as the atmosphere at work or at home. Play can serve as learning opportunities, make people and life more interesting and memorable, and serve to build better, more trusting relationships with others.[138]

There are many benefits of play in terms of mental, physical, and social benefits for both animals and humans, yet rarely do we make it a regular and intentional focus in our lives. I want to change that by bringing your attention to the power of play in this chapter.

## All Work and No Play?

Maybe it's just me, and perhaps it's because I am an empty nester now, but when I reflect back to my younger years and compare our lives today to those of my parents and their friends back forty or more years ago, it seems as if they enjoyed more fun and play in their adult lives than we do now. Why, I wonder? Is it because we have different priorities than they did years ago? Are our lives busier or more complicated than theirs were? Whatever the reason, I think we need to make play more of a priority now. Do you agree? (If not, I think reading this chapter will still be enlightening and educational for you. It may even change your mind about bringing more play into your life!)

Does it feel to you that we just work more and have less free time to play? You would think that with more technology and "time-saving" devices available to us now that we would be granted more available time to play, but that doesn't seem to be the case. At least not in my experience for the past two decades.

I'm going to suggest that we start prioritizing play. We need more play in our lives because we need to have fun and lighten up. We need to laugh more. We need to get together more with family and friends. We need more social connection and bonding.

The COVID pandemic drove us apart and isolated us into our homes, away from extended family and friends. And for many, it's still a lonely, depressing situation because they created new habits, staying inside, working from home perhaps, and having little socialization with friends and colleagues. For me, getting outside the home to network more with other like-minded businesspeople after Covid truly filled my soul. We are social creatures, so we need to be around other people. We need to connect with people even beyond our family. I encourage you to not only get out more and network with others, but to think consciously and purposefully of bringing more play and fun into your life . . . like our pets do, naturally.

## Animals and Social Play

Animals seek out play. They engage in social play with other animals (and humans) as well as in self-play because they enjoy it!

Social play for dogs is a voluntary activity that is directed toward another animal or human using actions that they use in other contexts. For instance, they will use actions like chasing, pouncing, mounting, humping, biting, and body slamming when playing that you may witness them doing if they were mating or fighting with another dog. You may see dogs start chasing each other, and other dogs joining in when in a leash-free park, for instance. Play is fun for them and highly contagious.

From studies of dogs interacting together, it has been noted that there are rules of play amongst them, and if they don't play fairly, they are avoided. The same goes for wild coyotes.

## The Rules of Play for Dogs

If you've ever witnessed a dog approach another dog (or yourself) with his forelegs on the ground and his rear end in the air, this is a bow. This is how they ask another dog (or human) if they want to play. Dogs want other dogs to play fairly and to follow the rules. If the other dogs don't play fairly or try to dominate or mate with them during 'play time,' they are avoided and become loners. In the wild, these "loner" dogs (as well as with wild coyotes) fail to form social bonds and actually suffer four times the mortality rate in their first year because they lose the support and benefits of a pack. Even in the wild, the lack of play reveals how detrimental it is to a living being.[139] (I would argue that for humans, too, the lack of play is detrimental to one's mental and emotional health. It can get pretty lonely and desolate when not incorporating fun and play into one's life and may lead to feelings of depression and melancholy.)

Just like with young children learning to play by the rules, so do dogs. As young puppies are learning to play, if they behave too roughly, the other, older dogs will let them know by growling, biting, or snapping for instance. The pup quickly learns what the acceptable boundaries are and how to play nice.

Here are the Four Basic Rules of Play for dogs:

1.  Ask first (bow)
2.  Be honest (play fairly)
3.  Follow the rules (it's not to mate nor dominate)
4.  Admit when you're wrong[140]

From fMRI studies done by Gregory Berns, a neuroscientist who studied the reactions of dogs, it was revealed that animals do, in fact, experience positive emotions. We also believe that from activities like play, animals also experience happiness, joy, and fun, just like we do.[141]

## Animals and Self-Play

Animals will also engage in self-play like chasing their tail, playing with balls or other objects on their own (e.g., throwing them up in the air), and playing tug of war with other animals or humans.[142] We figure they do so simply because it is fun and creates positive emotions.

## Cats and Play

It's hard to think of play without thinking of cats and kittens. They epitomize play. Their whole life seems to revolve around playing! If they're not chasing a shadow of a butterfly inside a sunny window, pawing at or flicking a dust bunny curiously around the floor as if playing in a hockey game, or whipping around the house at full speed looking wild and rabid, then they're either slowly snacking or soundly snoozing on the back of a couch or a warm comfy bed. Don't you wish you had the life of a cat?

We enjoy videos of cats on social media because their playful nature can get them into trouble. It's almost as if we enjoy living vicariously through them. Their curiosity makes them do funny things that we wish we had the guts to do. And while they are usually quite agile, they can get into some messy situations. Luckily for them, they are quick and fast on their feet, and are highly reactive and responsive, and can jump and bounce out of situations almost as fast as they got into them!

Like dogs, cats also abide by the rules of play. They will display certain behaviors that indicate to other pets or humans that they are interested in playing. They will do the similar crouch, like dogs, with their hind end up in the air, or they may crouch their entire body down close to the ground, moving their body back and forth as if revving up and getting ready to pounce. For cats, playing behavior is very much like hunting behavior–eyes focused intently on an object (or you), ears forward, body crouched, and stalking whatever they are going to attack. Then they pounce! As long as nails are in, body is relaxed, and they're not screeching or hissing, this is play and not fighting.

Play for cats is an indicator of happiness. They want to engage with other pets and humans when they're happy. Happy cats will purr, groom, and rub their head and body alongside you or a wall, leaving their scent as a mark. You may witness a cat kneading a blanket or your shirt, which indicates it feels content and safe. A cat may also chirp or meow to show that it wants to interact and play with you.[143]

## Be More Playful Like Your Cat

We need to be more playful like cats. While I'm not saying to rub and knead, I am suggesting that perhaps we speak up (chirp) more often and ask others to engage in a game of Scrabble or a fun game of touch football outside. Be spontaneous. Act crazy and goofy. Have fun. Even if for an hour or so on a Sunday afternoon. Get off the couch, step away from your devices, and go outside and have some fun.

Play is important for all ages. It can improve one's relationships, mood, and work atmosphere. It helps relieve stress, makes one more engaging and interesting, and is a great activity for connecting and building relationships. Through play, we learn to trust one another and feel safe.

Recently, my brother's wife reached out to me (via a text; a "pseudo-chirp") to ask if I wanted to go on a roller coaster ride together the next day. Now, this is a roller coaster ride through a ski mountain and a forest of trees, where you can choose how fast or slow you want to go through the manual use of a brake in your solitary coaster cart. My husband and I had ridden this coaster earlier in the summer and found that it is surprisingly fast with lots of bends and curves. I was nervous while my husband had a blast and of course went through it at top speed, not using the brake at all, while I, being the "scaredy-cat" (excuse the cat pun), rode the brake the entire ride! My first inclination was to simply decline the offer, as I had other tasks to do (like writing this chapter). However, in wanting to practice what I preach and institute more fun into my life, too, I accepted, and we went to the mountain. I failed to tell you at the beginning of this story however, that we are all in our fifties, it was a cold, drizzly day, and over an hour car drive away. But that did not stop us. We went, excited as little kids, to ride a roller coaster in late fall in Central Ontario.

Unfortunately, because it was raining, the roller coaster was "delayed" or not running at that time (probably a good thing, because I could just visualize my cart going careening off the railing and landing atop some pine trees, even though I was driving the slowest!). Nevertheless, we enjoyed each other's company, walking and shopping at the outdoor ski resort village, dancing in and around the puddles to the outdoor music, and dining at the local Irish pub. We had lots of laughs, created some wonderful memories, and are planning to try to ride the coaster at another time soon. (Truthfully, I'm hoping for next summer when it's warmer.) I tell this story to point out that you're never too old to play, and even though the conditions may not be perfect, do it anyway. It can still be fun. Fun is what you make of it.

## Have a Playful Attitude

Having a playful attitude can also keep your relationship with your spouse fresh and exciting! Plan a date night one night a month or week. Go out for dinner or to a movie. Or make a new meal together and set a fancy table with the "good set" of dishes and dress up formally for it. Pack a picnic and go for a hike to a park one summer day or book a winter getaway and go snowshoeing together. Planning fun activities to enjoy together will not only strengthen the bond between you and your spouse and allow for some stress relief in your week, but also make for some fond memories to cherish for years to come.

A playful attitude can also attract family and friends to you because you're seen as fun and engaging to be around. Few people want to be around someone who is depressing or negative all the time. A happy, playful attitude can brighten any room, bringing energy and vitality to it as well as to others–all of which are beneficial whether at work or home.

## Benefits of Play

There are lots of health benefits that result from play, as well.

Play can:

- Relieve stress
- Release positive endorphins
- Promote an overall sense of well-being
- Improve brain function
- Reduce depression
- Provide better learning opportunities
- Improve resistance to disease
- Increase the amount of exercise you get

That's an incredible list of positive and healthy benefits. Overall, play helps you function at your best! So why wouldn't we all want to bring more play into our lives?

## Play at Work

By incorporating more play at work (and home), our lives become more enjoyable, lighter, and more memorable. Play improves our social skills too, giving us more opportunities to create friends and long-lasting, meaningful relationships, which aid in our overall emotional well-being. And when you're feeling good about yourself and your life, your work improves, too.

Play can be brought into the workplace in many ways. Staff lunchrooms can have card and board games, encouraging workers to play together at break time or on their lunch period. A basketball net outside or in a workplace gym can encourage a group of workers to shoot some hoops or play a game of basketball on their lunch hour. Expending energy like this reduces stress, improves cardiovascular health, and creates social bonds amongst the team players.

Using games or a puzzle format in the boardroom is an interesting way to tap into the creative side of workers' brains, allowing for new insights and possible solutions to come forth.

## Play at Home

Creating opportunities to play at home, like hosting a family game night or organizing some fun outdoor activities for the family, can bring the family together and create wonderful memories to be cherished for years. For several years in a row, our immediate and extended family, including cousins, aunts, and uncles would come together, form teams, and compete against each other at our summer cottage. We set up a number of outdoor land games as well as kayak and hip-wader races in the lake that were so much fun that they became the highlight of every summer for many of my family members. These weekends also became some of my most favorite cottage memories.

## Play with Your Children

Making time to play more with your children will get you and them more active, show your kids a "different side of you" as their parent, strengthen your relationship with your kids, and build great memories together.

Be playful. Chase your kids around the house. Tickle or play tackle with them outside in the yard. Rake the leaves and jump in the leaf piles with them. Be like a kid yourself! You will not only happily surprise your kids acting like *one of them*, but you'll have fun too. Chances are they will want to be around you more because you'll be seen as fun. This is the essence of enjoying each other. Incorporating more fun and play into our lives. Share a hearty laugh together, be silly. Do it daily if possible. It's wonderful for the soul.

By being more playful and sharing in some goofiness with your kids, they learn that it's okay to be expressive. Refrain from harboring your feelings and emotions inside or trying to stay poised, especially for your family. Show your kids that it's good to express yourself, whether happy or sad. This openness will hopefully allow your own kids to freely express themselves and their harbored feelings to you, the parent, when they have witnessed seeing you express your feelings and honesty with them. Let them know that it is totally fine and acceptable to release and show their emotions or to act goofy crazy once in a while (or regularly).[144]

## Play with Your Pet

Don't forget to play with your pet! Let loose and act like a kid with your pet. Get down on the ground and play tug of war with a stuffed toy with your dog. I am sure you will enjoy it as much as your dog will enjoy the attention you are showing him or her. Laugh at your cat playing chase with a fake mouse or a laser pointer. Not only will you enjoy some play time and bonding time with your cat, but the extra exercise will do your cat good, especially if it's an indoor cat. In an article in *Science Direct*, Marc Bekoff explains that, "When we study play and fun, we're really studying ourselves as well. And, it's fun to do. It's a win-win for all involved."[145]

## Adult Playfulness

We also see play and playfulness trigger positive emotions in people. Neurotransmitters like dopamine and oxytocin are released during episodes of play. Dopamine is released when we like what we are doing, it's pleasurable, or a reward is involved. Oxytocin is released with social bonding activities like in team events and activities.

A study by Proyer in the *European Journal of Humor Research* showed that playfulness in adults, as a state of mind, also has several benefits (in addition to the ones listed earlier from play) such as:

o   Increased motivation

o   Increased creativity

o   Increased spontaneity

- Increased coordination and strength

- Improved cardiovascular health and physical fitness

- A positive attitude about the workplace and/or increased job satisfaction

- A greater quality of life

Overall, adult playfulness relates positively to life satisfaction. Someone who is more playful will be more inclined to be more active and engage in more enjoyable activities.[146]

So, as you can see, there are lots of good reasons to bring more play into your life–social, physical, and emotional, all resulting in a better sense of well-being. There are no downsides to incorporating more play into your life, other than taking the time to do so. But the time invested would be well worth it.

## Be Playful about You

Speaking of playfulness, you don't have to just limit it to behaviors and actions. Be playful with your looks, too! Go out and get a new outfit, perhaps a bit more "hip" or daring than usual. Get your hair cut in a new style. Try a drastic hair color change. Shake it up a bit. Step outside of your comfort zone here too. It can be fun for you and others. Do these as one of your mini-rewards, perhaps, in your Leap Plan. Earn it. Get excited about it and work for it. Feel the anticipation and have fun with it.

I did this years ago with my hair. My natural hair color is a medium brown color. The first time I had my hair cut short and colored to an almost platinum blonde color, I called my husband to tell him. Boy, was he eager to come home to see for himself the new "platinum blonde" woman he was married to. He loved it! Other women loved it, too, and wished they had the guts to do the same. (A few did ask who my stylist was as they were inspired to make a change and have some fun with their looks, as well.)

If you feel inspired to do it too, then do it. You may feel renewed, like a different person. Perhaps the physical change you see in yourself may even spark new positive behaviors in you as well, like wanting to get out and socialize more, or hit the gym. And heck, if you don't like the new color of your hair, it will grow back to its original color in time, or the stylist can change it back. But go ahead and have fun with it. Chances are, you may find a new look that really works for you and makes you feel great and even more confident. And isn't that what it's all about–making you feel good about yourself?

## Laughter Is the Best Medicine

As I mentioned earlier, a big advantage of play is having and sharing a hearty laugh with someone else. There is so much good that laughter can bring. It's therapeutic, it makes one feel good, and it brings a lot of emotional benefits as well as physical benefits.

Much has been written about the benefits of laughter. It's even good for your heart. A recent 2023 study at the Hospital de Clínicas de Porto Alegre in Brazil found that laughter releases endorphins, which reduces inflammation and helps the heart and blood vessels relax, allowing them to function better. Laughter was also shown to reduce the levels of stress hormones, which can place strain on the heart, and thus can decrease the risk of heart attack and stroke! So, the old saying that "laughter is the best medicine" holds true in many ways for us–emotionally, physically, for our soul and . . . our heart.[147]

I hope I have proven to you why you need to incorporate more play into your life, if not daily, at least regularly. It will do *you* good as well as those around you. It will benefit your work and home life.

I give you permission to act and behave like a kid again. Play, laugh, and act carefree. You will love it and so will others around you, including your pet! You will have a more playful spirit, and will attract and be surrounded by other positive, jovial people who want to join you and learn how to have more fun in the process. You will be healthier, less stressed, have a healthier perspective on life and view life in a whole new light.

So, get out there and play!

## FETCH STEPS

Make a point of playing with your pet each day. If it's the same time each day, all the better, as it will form a habit for both of you. Your pet will look forward to this time together and you will likely remember to do it.

1. Be a kid for an hour or two this week. Tickle your child, rake and jump in a pile of leaves with your family, or chase your child around the house. Have fun and enjoy some laughter together. The rewards will be so worth it.

2. Plan a family game night or a date night for this coming weekend. See how much more rested you feel for the next day at work.

3. Look for fun ways to use your gift. Not only will you enjoy using your gift, but you and others will benefit from you having a playful attitude about sharing it!

# 18

# Celebrate . . . Then Chill Like Your Cat

**Congratulations! You've completed your Leap Plan** and accomplished your end goal: your Stretch goal.

How do you feel? Do you feel accomplished, exuberant, strong, and confident? Has it sunk in yet that you're done? Was it easier to reach your goal than you had originally thought? By staying committed to, and trusting the process in following your Leap Plan, was it easy to stay focused?

I hope you are rewarding yourself as you had suggested in your plan. Do not forego this step! This chapter is here solely to remind you to reap the major reward that you set out in your Leap Plan for accomplishing your Stretch goal. Take time to revel in it, celebrate it, reflect on your journey, and rest. Because you've earned it.

## Reap Your Reward

Take the time to reap your reward. Let it sink in. Journal how you feel right now (I delve deeper into this idea later in the chapter). Make this feeling last. You will want to come back and read this again in the future (probably more than once) as potential motivation and proof to yourself that if you did it once, you know you can do it again!

Tell others what you just did. Call up a family member or a friend and tell them. By sharing your big WIN, you get to re-live those wonderful feelings all over again. Better yet, go out and celebrate together. Make it a big deal–because it is.

Take this time to revel in, and feel the joy from, what you have accomplished here.

Share and celebrate your win with your helpers of your pack, your coach (if you used one), and your accountability partner(s). Have a "Pack Party." Make this celebration last. Fill your soul with lots of positive, heartfelt emotions that will create a lasting memory from which to draw upon for years to come. It may even become one of the highlights of your life!

If you didn't accomplish your goal yet, don't fret. Perhaps you just need more time to do so. Maybe there's something you still need to learn in order to finish your Leap Plan. Perhaps you need to adjust your goal somewhat. Whatever the reason(s), it's fine. Like I said earlier in the book, this is not a race. It's your journey. Use the process to fit you, your schedule, and your lifestyle. No one's life is the same. We are all unique–and that is what's wonderful about being human. As long as you are making progress, it's all good. Celebrate the progress that you have made so far. Reach out to your helpers or coach and see where to adjust your Leap Plan, if needed. I want you to win here.

Just don't give up on your Stretch goal, and most certainly, do not give up on yourself. Keep going. Remember the old saying, "Rome wasn't built in a day." Celebrate your progress, take a little break, and keep on truckin' with following your Leap Plan. Surround yourself with your pack for motivation and support. You can do this.

## Celebrate–It's Good for You!

Why am I making such a fuss about getting you to celebrate? Because just like with rewards and play, there are many benefits that I want you to experience from celebrating.

Celebrating boosts your confidence and well-being. By proving to yourself that you can do what you envision and set your mind to, you create a stronger belief in, and feel good about, yourself.[148] With greater self-esteem, you can continue to do great things and become more, if you wish. It sets in motion a momentum of believing in yourself, stretching you more, and accomplishing more. With such momentum, who knows where this confidence train will take you? I would suggest you hop on it, ride it, and see where it takes you. It could just be the ride of your life! Have fun and enjoy it.

Alongside the confidence and momentum that you gain by celebrating your wins, celebrating also gives you motivation to seek more (and likely bigger) goals.[149] You start developing a growth mindset where you get excited not only about the goals you are going to accomplish, but more importantly, who you will become in the process. The person you've always strived to become–the person with a heartfelt purpose who is juiced by life and living with joy and passion.

Like a snowball rolling down the hill, as you achieve your goals and build your confidence and momentum, the size of your goals gets bigger, along with the list of your accomplishments and the perception of yourself. Your growth mindset has now evolved into a success mindset, where you believe that you can do what you set your mind to do. Now you see yourself as 'successful' and not just 'trying to be successful.'

Celebrating success helps you overcome our natural tendency as humans to focus on the negatives. Called negative bias, your brain seems to like to focus on problems and failures (perhaps for survival reasons years ago), so by deliberately celebrating success, you shift your brain to a positive focus. When you turn your attention to the feelings of success, you get a sense of satisfaction and pride (internal rewards, so to speak). It's these feelings that will bring you more motivation and belief to set more goals and achieve more wins and success in the future.[150] Even if your future goals don't all lead to success, you still have won because you've tried, you've experienced, you've learned, and you've grown as a person–all good things for your personal development, mindset, and perspective.

And of course, the best reason for celebrating the achievement of your goal is that it makes you feel good.[151] When you celebrate, whether by dancing, singing, partying with others, or by simply getting your major reward and writing in your journal, this process makes you feel happier, elated, and better overall.[152] And isn't feeling happier about yourself what you want in life?

If we don't recognize our accomplishments in life, even daily wins (through mini-rewards), we run the risk of feeling like life is one big never-ending goal. If we never celebrate our wins, even small ones, we fail to experience the positive emotions that come from celebrating. Life becomes like a chore, and we become almost robotic with no emotion. What fun is there in that? We are sentient beings who need to feel emotions. The wins, even the small ones, build our self-esteem and hopefulness and let us know that we can get things done, that we are worthwhile, and that we deserve to have fun celebrating them.

So always celebrate your accomplishments, no matter how big or small. It will do you a world of good.

## Rest

Once you've reaped your reward and celebrated, now I want you to rest. Rest like a cat. Take time to chill. Not only have you earned a good rest, it's also very important to rest.

Resting is required to recharge.

When I think of resting, I think of a cat. When they're not playing or eating, they're sleeping. Cats sleep to recharge. They seem to sleep most of the time–and they do. Cats sleep on average fifteen hours a day; some up to twenty hours a day. Most of this time is spent in an almost-waking rest state in which their senses of smell and hearing are still "on." Heaven forbid if they were to miss a bird or a mouse nearby! In this almost-waking state, cats are in a type of sleep called non-REM (Rapid Eye Movement), in which the body still gets to recharge but also conserves energy. Cats do, however, undergo a deeper REM sleep for about 25 percent of their total sleeping time, which helps to regulate emotions (yes, believe it or not, cats do have emotions) and other recovery processes.[153]

With all that sleep and energy being conserved over fifteen hours, it's no wonder that cats have so much energy when they finally wake up. Energy to pounce, stalk, chase, and run wild around the house–all the things we love about cats. If only I could wake up like that . . .

Resting can have similar benefits for you. It increases your energy levels, makes you more productive, improves your focus, and lowers your stress levels. Resting improves your mood, too. When you relax, you feel happier. You feel more positive towards things and are more capable of handling challenges.[154]

Rest and relaxation can also provide you with a greater sense of fulfillment and satisfaction in life, which is what we are all aiming to achieve here. So, you see, cats have it right. Just make sure to get those cat naps in!

However, there are other ways to rest and relax than catnapping or sleeping. And no, I'm not suggesting that you sleep for fifteen hours a day like a cat. That simply wouldn't be conducive to being productive, now, would it? Besides, you wouldn't have time to use, hone, and share your gift, which is the goal of this book and hopefully yours, too!

Here are some other ways to relax:

- Exercise

- Meditation and mindfulness

- Yoga

- Reading

- Taking a vacation

- Spending time with family and friends[155]

Yet again, exercise is the answer to bringing so many good things into our lives. It's also one of the most effective forms of relaxation, believe it or not. Exercise actively rests your mind but not your body through the release of many neurotransmitters and brain chemicals such as dopamine, serotonin, norepinephrine, endocannabinoids, and BDNF (brain-derived neurotropic factor). The latter two chemical types are what give runners the 'runner's high' as well as reducing pain.[156] For me, running makes me feel good, improves my cardiovascular fitness, and allows me to clear my head. I can think better, more clearly, and more creatively during and after a good thirty-minute run.

So much has been written about the many benefits of meditation and mindfulness for relaxing the mind that I'm not going to delve into it here. Just know that many swear by it to rest and relax the mind, lower stress and anxiety, improve focus, and strengthen the immune system. [157]

Yoga is great for stretching the body as well as relaxing the mind. There are many different types of yoga, as well. Take a class at your local gym or community center and try it out for yourself.

Reading is a great way to relax with the added benefit of either learning (if a non-fiction or self-help book) or getting lost in another world (fiction book). I am a voracious reader and find this to be the best form of true relaxation for me.

Of course, I am always up for taking a vacation to relax! Getting away to a new environment to explore a different culture, try new foods, and engage in new adventures is like play. There are inherent lessons that abound when you travel, like learning how others live and communicate as well as learning about yourself. Whenever I have gone on a vacation, not only do I gain a sense of appreciation for where I live, but I often find I get some of my best, most creative business ideas when I am traveling. Probably because my mind is rested, almost cleared (think blank slate), not bombarded or interrupted by our normal daily tasks and devices, and I have time to think. (I think I need more vacations . . . wink wink.)

Spending time with family and friends where you can laugh together, be yourself, 'break bread,' play games, or just talk and hang out is a great way to relax, fill your soul, and create wonderful memories to cherish forever.

So, take time to rest. You've earned it. Rest your mind and your body, because it's essential for a healthy, balanced life. You and I are simply not meant to go full steam all the time.

## Journal to Reflect

Once you've 'come down' from the fun of your "Pack Party" and had a good rest, this would be a good time to journal. While the feelings are still fresh and somewhat present, journal what it feels like to have accomplished your big, hairy, audacious goal. Express your feelings on paper (if you're writing in a physical journal). You could also choose to write your feelings of accomplishment digitally, like in a blog online, or create a video of yourself speaking about it. You choose which form you prefer to express your journey and feelings in, both for yourself and perhaps others.

In your journal, write about the path you took and the hurdles you faced. Explain how you overcame or dealt with them. Reveal the lessons that you learned along your journey. Remind yourself of all the things you did well. Don't focus on the negatives. Getting the details and emotions on paper allows you to remember them. Taking this time to reflect and acknowledge how far you've come and how you may have changed will not only be very revealing for you, but also for others (i.e. future generations of your family, for instance) who get to read about your journey. Your story here could be the necessary fuel to motivate and guide others who want to do something similar in their own lives. The steps of your journey could help others walk a similar path. In essence, your journal is part of the legacy you leave behind.

I have been journaling for over thirty years. I have a large box of journals, because my personal development library is filled with other writers' books. It's fun to look back and read what my goals were ten, twenty, or thirty years ago. Some of those goals were never achieved, but luckily many were. I could see trends in my thinking and noticed how much I have grown just from reading my journals over time. Reading about my values from years ago revealed to me that those haven't really changed much over the years. Similarly, my overall life mission statement of wanting to inspire others to be their best, bring value wherever I go, and to make an impact, hasn't changed much over the years either. I hope I have succeeded here in inspiring you to be your best and make an impact for others.

## Self-Reflection

In this fast-paced world where we feel we have to keep working, checking off the many tasks on our daily list (except the ones on your Leap Plan, of course), and then doing it all again the next day, we often forget to stop and acknowledge all that we have done. We don't take time to reflect and see how far we have come, and this is a big miss. We won't truly live if we don't stop to acknowledge what we've accomplished. We'll miss out on the joys of life, the lessons learned from the fruits of our labor, and the internal feelings of growth, satisfaction, and pride in reaching our goals. Be grateful for what you've been able to do here in your Leap Plan.

Taking time to self-reflect and think about the journey you just embarked on will force your brain to not only rest but also to remember the experiences, the challenges, the meaning, the emotions felt, and find the lessons learned. This information helps you grow as a person and could even show you how to guide others that may be going through something similar or who may have a similar goal down the road. Hopefully, you will discover some valuable insights and maybe even a breakthrough that you weren't expecting (what a fun surprise reward that would be!).

As you're reflecting on this process, here are some thought-provoking questions to ask yourself:

1. What did I learn from that?
2. Could I have done better at it?
3. Was that the best approach to do "X"?
4. What would I have done differently, if anything, if I were to do it again?
5. Knowing what I know now, what is the next logical step or goal for me?

Think about these questions or others that you may come up with while sitting quietly alone for about ten to thirty minutes, or going for a walk, or talking to a colleague or your coach about them. Journal your answers. Do it daily for a week or so. You may have a different perspective from one day to the next (depending on how much rest your brain has had). Put careful thought into answering these questions. Reflecting helps you to develop your skills and review your effectiveness. By learning from what you did and experienced, you grow.[158] You become more self-aware of who you are and what you're capable of.

By taking time for self-reflection, you actually save time and energy moving forward because you learn to focus on the things that truly matter to you, your progress, and your life.

As Peter Drucker once said, "Follow effective action with quiet reflection. From the quiet reflection, will come even more effective action."[159]

Reflection requires courage and being honest with yourself. It's taking a hard look at yourself, what and how you did something, what you learned, and how you could do better.[160] It's self-analysis and self-improvement all in one.

Even if we do something wrong, we can still learn from it. I would argue (as many would) that often it's the mistakes or failures from which we learn the most, perhaps because they instill more emotion in us and are thus better remembered.

All in all, the important point is to learn from your experiences and grow. Grow into becoming the person you desire to be. Grow to help others; maybe even to lead others. Lead or teach from your experiences, learnings, and perspective . . . just like I am doing here in this book.

For some people, this book may be odd in that I am trying to help people by using analogies and learnings from animals, but this idea came from *my* unique perspective as a veterinarian and pet owner. This idea also stems from a belief (true or not, who knows) of mine that life or the universe leaves clues. Nature, as part of the universe, for me, seems to hold a lot of answers. Animals, like our pets, are part of that natural world. So why not learn from them? They could hold some of those clues from which we can learn; learn how to do things more easily and simply, and in flow, like in nature.

Combining simple animal behavior techniques and more natural ways of doing things to help people simply and more easily change habits and behaviors from "bad to good," or start a new one, or learn how to believe in themselves (a big lack I witnessed in many), could I help many of those who didn't know how or where to start or who didn't think they *could* change? If I could show them what is possible using a simple formula like Dig. Leap. Play. that we see used with, and by, animals, could they believe in themselves to do it, too? My hope is yes.

If I could show others what I was able to do and accomplish in my life so far, often with little to no background and not believing that I was anything special, then perhaps I could get you to believe that it's possible for you too to accomplish what you want.

Now that I am entering the "fall" of my life, I felt it was time to share my learnings, experiences, and perspective here in this book (and my course of the same name) in hopes of helping others from this unique perspective of combining animal behavior and personal development. It may not make sense for people who aren't pet lovers or owners, but for those who are, and who can relate to and love animals, this book is a refreshing change in the personal development space.

The ultimate goal (and test), of course, is that this book made you think. Made you think about what is possible for you; made you think bigger and made you believe that you *can* do it.

The main reason I wrote this book was to get you to take action. If you don't take action steps toward your goals and what you want out of life, then chances are pretty good that you won't arrive there.

Again, if I can show you a simple path to get there that's fun and manageable, and reveal how others, including animals, could make it happen, perhaps you will take the necessary action steps to achieve your goals. In doing so, you could see for yourself that it's not as hard as you think, that you are capable, and you are special. Taking it one step further and having you focus on finding and using your gift, my hope here is that the action you take will be easier and more fulfilling for you because it involves using your gift. As I mentioned in the beginning, I believe that we all have a gift and our mission in life is to find, hone, and use it for the betterment of others.

By reflecting on your journey, experiences, and learning, and continuing to hone your gift, you'll be better able to serve others and feel even more joy and personal fulfillment.

I'll say this again to you, "I am proud of you."

I am proud of you for picking up this book, embracing my unique perspective and analogies, following through on the exercises, and making it to the end. I hope you committed to the process and were successful in achieving your Stretch goal. Congratulations from the bottom of my heart for doing so. You should be proud of yourself, too!

You have also brought me great joy and fulfillment by reaching your goal and feeling happier and more confident about yourself (as you should). Now I feel my "gift" has served others . . . YOU . . . and that makes me feel complete in making my goal of wanting to make an impact on others.

I had so much fun and learned so much while writing this book, that I plan on writing more books about using this same analogy of learning from pets. I plan to write about how you can more specifically take the leap for better health, for building your dream business, for speaking up for yourself, and for getting into tip-top fitness shape. I hope you will elect to join me on one or all of these journeys, too.

Stay tuned . . .

I couldn't finish this chapter without leaving you with a few more Fetch Steps to complete.

# FETCH STEPS

1.  Share your recent journey with others, whether family, friends, or online.

2.  Enroll in my course to get even more out of this process. For details, check out my website: https://www.drteresa-woolard.com

3.  Decide on your next Stretch goal and write a new Leap Plan.


THE END.

# What's Next?

Now that you've celebrated, shared, rested, and reflected on your BIG accomplishment, I want to encourage you to build upon and keep this momentum going.

Start thinking about what you can do next. Where can you take your gift, learnings, and new bigger thinking now?

Take some time, dig deep, and use your intuition (your animal instinct). Listen to what your body and heart are telling you to do. Trust yourself. Are some ideas popping into your head? Be sure to jot them down in your *Dig. Leap. Play. Companion Workbook* (available at companionworkbook.com). Bounce them off your coach, colleagues, friends, and family members.

Have fun brainstorming. Explore some of the ideas and really research them. If you feel a tang of excitement with one or two of them, go deeper into them. It just may be the universe leaving one of its clues for you. Leverage and trust this inspired action. In my opinion and experience, this is the juice of life. Getting excited and moved by what inspires you, coupled with knowing what and how to use your gift, is GOLD. This is where and how you should be living.

Go for it!

Shine and share your gift for all the world to see (if you want). Most importantly, do it for yourself.

Live the life you were meant to live.

Be the person you were meant to be.

Live life with more joy and fulfillment, my friend.

Now go play with your pet!

# Additional Resources

**The Dig. Leap. Play. concept goes beyond** this book to include additional, helpful resources through my website www.drteresawoolard. com, blogs, courses, and seminars. My mission is to provide a wealth of information and tools to help you further shape your mindset, develop your gift, build on your confidence, and help you take the leap to achieving more of your dream goals and live the life you desire. I hope you will take advantage of some or all of these tools, as well.

# Acknowledgments

**This book has taken over ten years** to come to fruition and was re-ignited from a prompting that came from within me (there's that self-intuition knocking at my door again). It has been a project of the heart that I knew I just had to do, otherwise I would have regretted it later in life had I not. I wrote this book for me, my Mom (who was the first reader and cheerleader for this book), but most importantly, I did it for you. I hope it has helped you realize your gift once and for all and has prompted you to take the leap to doing something with it, so you too don't experience regret later in life.

Taking the necessary time to experience and learn other skills before re-writing this book was well worth it. It became a much better book; better than I ever would have imagined. But I didn't do it alone. I had a team of supporters and coaches who assisted me in the production of this book that I would like to thank.

Thank you to Dr. Sarah E. Brown, my Get Published Now (GPN) book coach, for being so enthusiastic, encouraging, and helpful throughout the re-writing of this book and for keeping me on task. Thank you also to the other GPN coaches Steve Harrison, Cristina Smith, and Debby Englander for their wisdom and knowledge, and a special thank you to my talented editor Valerie Costa for helping to create such a polished piece.

To Christy Day, of Constellation Book Services, and graphic designer extraordinaire, thank you from the bottom of my heart. You have created a book that I am in LOVE with and proud to share!

Thank you to my beta readers Taylor Harvie, Hana Erickson, Cher Cunningham, and Mark Woolard. I greatly appreciated your feedback in the early development of *Dig. Leap. Play.*

A special thank you to my husband, Mark Woolard, for your unwavering and ever-lasting patience, support, and encouragement, as well as for allowing me the time, space, and freedom to rewrite this book. Your creative wisdom has also been appreciated in the marketing of not only this book but its ancillary components.

# References

**CHAPTER 1**

1   Gift definition and meaning | Collins English dictionary. Accessed February 1, 2024. https://www.collinsdictionary.com/dictionary/english/gift.

2   "Gift Definition & Meaning." Merriam-Webster. Accessed January 31, 2024. https://www.merriam-webster.com/dictionary/gift.

3   "Watch Arnold: Netflix Official Site." Watch Arnold | Netflix Official Site, June 7, 2023. https://www.netflix.com/title/81317673.

4   ibid

5   Ware, Bronnie. The top five regrets of the dying: A life transformed by the dearly departing. London: Hay House Inc., 2012.

6   Taylor, Adam. "The Work Life Equation." LinkedIn, May 21, 2019. https://www.linkedin.com/pulse/work-life-equation-understanding-your-talents-gifts-ksas-taylor-ccp/.

7   Allianz Life Insurance Company of North America. The Gift of Time-The Allianz Longevity Project: How Americans are approaching the prospect of a longer life, 2016. https://www.allianzlife.com/~/media/files/global/documents/2016/07/21/15/57/ent-1848-n.pdf

8   Ibid.

**CHAPTER 2**

9   Rehman, Ibraheem. "Classical Conditioning." U.S. National Library of Medicine, August 14, 2023. https://www.ncbi.nlm.nih.gov/books/NBK470326/#:~:text=The%20dogs%20salivating%20for%20food,the%20conditioned%20response%20was%20salivation., n.d.

10   Yin, Sophia A. Low stress handling, restraint and behavior modification of Dogs & Cats: Techniques for developing patients who love their visits. Davis, CA: CattleDog Pub., 2009.

### CHAPTER 3

11    Staff. "How to Boost Feel-Good Hormones Naturally." Henry Ford Health - Detroit, MI, May 3, 2021. https://www.henryford.com/blog/2021/05/how-to-boost-feel-good-hormones-naturally#:~:text=Dopamine%3A%20Often%20called%20the%20%22happy,Praised%20on%20the%20job%3F.

### CHAPTER 4

12    Peirce, Penney. The intuitive way: The definitive guide to increasing your awareness. New York, NY: Atria Paperback, 2009.

13    Socrates. "A Quote by Socrates." Goodreads. Accessed March 12, 2024. https://www.goodreads.com/quotes/452128-to-know-thyself-is-the-beginning-of-wisdom#:~:text=Quote%20by%20Socrates%3A%20%E2%80%9CTo%20know,is%20the%20beginning%20of%20wisdom.%E2%80%9D.

14    "A dog is at its happiest doing what its instincts tell him to do." Author and source unknown.

15    Ibid.

16    Quote, Quotable. "A Quote by Steve Jobs." Goodreads.com, Goodreads, Inc. Accessed February 1, 2024. https://www.goodreads.com/quotes/903982-you-ve-got-to-find-what-you-love-and-that-is.

17    Gawain, Shakti. Developing intuition: Practical guidance for daily life. Novato, CA: Nataraj Pub., 2002. Pgs. xi-xii

18    Peirce, Penney. The intuitive way: The definitive guide to increasing your awareness. New York, NY: Atria Paperback, 2009.

19    Reeds, Tiffany. "Solar Plexus Exercise & How It'll Help Boost Your Intuition." Inner Workings Of The Soul, July 29, 2021. https://www.innerworkingsofthesoul.com/new-blog/solar-plexus-exercise-amp-how-itll-help-boost-your-claircognizanceclairsentience.

20    Airth, Maria. "Solar Plexus Anatomy, Location & Function - Lesson | Study.Com." Study.com, November 23, 2023. https://study.com/academy/lesson/what-is-the-solar-plexus-definition-function-location.html.

21    Wolfgang von Goethe, Johann. "Johann Wolfgang von Goethequote Citation." AllAuthor. Accessed March 12, 2024. https://allauthor.com/cite/42097/.

22    Jobs, Steve. "A Quote by Steve Jobs." Goodreads. Accessed March 12, 2024. https://www.goodreads.com/quotes/445286-have-the-courage-to-follow-your-heart-and-intuition-they.

23    Peirce, Penney. The intuitive way: The definitive guide to increasing your awareness. New York, NY: Atria Paperback, 2009.

24    Tracy, Brian. Focal point: A proven system to simplify your life, double your productivity, and achieve all your goals. New York: AMACOM, 2005.


CHAPTER 6

25    Ziglar, Zig. "Zig Ziglar Quotes - Brainyquote." BrainyQuote.com. Accessed March 12, 2024. https://www.brainyquote.com/authors/zig-ziglar-quotes.


CHAPTER 7

26    "Mindset Definition & Meaning." Merriam-Webster. Accessed February 1, 2024. https://www.merriam-webster.com/dictionary/mindset.

27    Pettit, Mark. "5 Ways to Develop an Attitude of Gratitude - Lucemi Consulting." Lucemi Consulting: Productivity and Time Management Coach, April 10, 2021. https://lucemiconsulting.co.uk/attitude-of-gratitude/.


CHAPTER 8

28    Assaraf, John, and Murray Smith. The Answer: Grow any business, achieve financial freedom, and live an extraordinary life. New York, NY: Atria Paperback, N.S.W.: Simon & Schuster Inc, 2008.

29    Meyer, Karman. Eat to sleep: What to eat and when to eat it for a good night's sleep--every night. New York, NY: Adams Media, 2019. Pg 12-25

30    Ibid.


CHAPTER 9

31    Quicken. "How to Build Good Financial Habits: Quicken." Quicken Blog, May 15, 2020. https://www.quicken.com/blog/build-good-financial-habits/.

32    Duhigg, Charles. "Safety First, Benefits Follow." The Power of Habits, 2021.

33    Gladwell, Malcolm. Outliers. Harmondsworth, London: Penguin, 2009. Pg. 35-41.

34    Clear, James. Atomic habits: The life-changing million copy bestseller. Random House, 2018. Pg. 110-111

35    Philippa Lally, Cornelia H.M. van Jaarsveld, Henry W.W. Potts, and Jane Warde, "How are habits formed: Modeling habit formation in the real world," European Journal of Social Psychology. July 2009. https://onlinelibrary.wiley.com/doi/10.1002/ejsp.674

36    Roomer, Jari. "Science Says You Need Systems More than Willpower to Achieve Your Goals & Change Your Habits." Medium, July 17, 2019. https://medium.com/personal-growth-lab/science-says-you-need-systems-more-than-willpower-to-achieve-your-goals-change-your-habits-80274d3c0b70.

37    Markham Heid, "The Right Way to Change Your Habits," The Power of Habits, 2021, Pg. 68-71.


CHAPTER 10

38    Fogg, B. J. Tiny Habits: Why starting small makes Lasting Change Easy. London, England: Virgin Books, 2020.


CHAPTER 11

39    Georgiev, Deyan. "How Much Time Do People Spend on Social Media in 2024?" Techjury, January 3, 2024. https://techjury.net/blog/time-spent-on-social-media/#:~:text=An%20average%20user%20spends%202,is%20reserved%20for%20social%20media.

40    Becker, Joshua. "What to Do When a Distraction Becomes a Lifestyle." Becoming Minimalist, June 21, 2017. https://www.becomingminimalist.com/distractions/#:~:text=Distractions%20in%20life%20are%20not%20unique.

41    Smith, Jason. "Distractions: Internal & External Challenges That Derail Our Focus." LinkedIn, August 7, 2023. https://www.linkedin.com/pulse/distractions-internal-external-challenges-derail-our-focus-smith/.

42    "Distractions: Internal & External Challenges That Derail Our Focus." The Open University. Accessed February 1, 2024. https://help.open.ac.uk/distractions-and-procrastination#:~:text=Distractions%20can%20be%20real%20(e.g.,%2C%20say%2C%20writing%20an%20assignment.

43    Schatz, Itimar. "Solving Procrastination." Solving Procrastination. Accessed February 1, 2024. https://solvingprocrastination.com/why-people-procrastinate/#:~:text=The%20following%20are%20the%20key%20reasons%20people%20procrastinate%3A&text=Task%20aversiveness%20(i.e.%2C%20thinking%20a,it's%20unclear%20where%20to%20start).

44    Belling, Sherry. "4 Reasons Why Do Horses Wear Blinders." Horse Riding Guide, April 14, 2022. https://www.deephollowranch.com/why-do-horses-wear-blinders/.

45    "Blinkers (Horse Tack)." Wikipedia, October 29, 2023. https://en.wikipedia.org/wiki/Blinkers_(horse_tack)#:~:text=Many%20racehorse%20trainers%20believe%20that,especially%20on%20crowded%20city%20streets..

46    "Wear Space." Panasonic. Accessed February 1, 2024. https://panasonic.net/design/flf/works/wear-space/#:~:text=WEAR%20SPACE%20is%20a%20wearable,controls%20your%20field%20of%20view.

47    "Why Multitasking Doesn't Work." Cleveland Clinic, November 27, 2023. https://health.clevelandclinic.org/science-clear-multitasking-doesnt-work/#:~:text=We're%20really%20wired%20to,doing%20two%20things%20at%20once.

48    Ibid.

49    Ibid.

50    Srishti. "Average Screen Time Statistics for 2023." Elite Content Marketer, October 20, 2023. https://elitecontentmarketer.com/screen-time-statistics/.

51    Ibid.

52    Stoll, Julia. "Hours of TV Watched per Week by Age in Canada 2023." Statista, November 27, 2023. https://www.statista.com/statistics/234311/weekly-time-spent-watching-tv-in-canada-by-age-group/.

53    Howarth, Josh. "25 Startling Social Media Addiction Statistics (2024)." Exploding Topics, November 29, 2023. https://explodingtopics.com/blog/social-media-addiction.

54    Brown, Lorna, and Daria J. Kuss. "Fear of Missing out, Mental Wellbeing, and Social Connectedness: A Seven-Day Social Media Abstinence Trial." MDPI, June 24, 2020. https://www.mdpi.com/1660-4601/17/12/4566.

55    Biscontini, Gianna. "'I Quit Social Media. I've Never Felt Happier.'" Newsweek, January 10, 2023. https://www.newsweek.com/quit-social-media-mental-health-benefits-1771907#:~:text=I've%20noticed%20I%20feel,every%20sense%20of%20the%20word.

56    "10 Focus Exercises to Help Improve Concentration Skills." Indeed, August 3, 2022. https://www.indeed.com/career-advice/career-development/focus-exercises.

57    Wooll, Maggie. "10 Concentration Exercises to Improve Your Focus." BetterUp, August 30, 2022. https://www.betterup.com/blog/concentration-exercises.

**CHAPTER 12**

58    https://www.steamboatsocceracademy.com/cani-constant-and-never-ending-improvement/#:~:text=It%20takes%20dedication%2C%20perseverance%2C%20and,Kaizen%20%3D%20change%20is%20good.

59    Kendra Cherry, MSEd. "How Brain Neurons Change over Time from Life Experience." Verywell Mind, November 8, 2022. https://www.verywellmind.com/what-is-brain-plasticity. .

60    Galván, Adriana. "Neural Plasticity of Development and Learning." Human brain mapping, April 27, 2010. https://www.ncbi.nlm.nih.gov/pmc/articles/PMC6871182/#:~:text=According%20to%20the%20theories%20of,and%20activity%E2%80%90dependent%20synaptic%20plasticity.

61    Cunnington, Ross, Jean-Claude Dreher, Joel B. Talcott, Donna Coch, David Bueno, Alejandro Maiche, María Castelló, and Vivian Reigosa Crespo. "Science of Learning Portal - Neuroplasticity: How the Brain Changes with Learning." IBE, June 28, 2021. https://sol-portal.ibe-unesco.org/articles/neuroplasticity-how-the-brain-changes-with-learning/#:~:text=Neuroplasticity%20is%20important%20for%20all,learning%20throughout%20all%20of%20life.

62    Sinha Dutta, Dr. Sanchari. "Hippocampus Functions." News Medical, August 21, 2019. https://www.news-medical.net/health/Hippocampus-Functions.aspx.

63    Ibid..

64    Ibid.

65    Maguire, Eleanor, Katherine Woollett, and Hugo Spiers. "London Taxi Drivers and Bus Drivers: A Structural MRI and Neuropsychological Analysis." NIH, 2006. https://pubmed.ncbi.nlm.nih.gov/17024677/.

66    Bernard, Sara. "Neuroplasticity: Learning Physically Changes the Brain." Edutopia, December 1, 2010. https://www.edutopia.org/neuroscience-brain-based-learning-neuroplasticity.

67    Ibid..

68    Cherry, Kendra. "How Brain Neurons Change over Time from Life Experience." Verywell Mind, November 8, 2022. https://www.verywellmind.com/what-is-brain-plasticity-2794886#:~:text=Neuroplasticity%20is%20the%20brain's%20ability,structural%20changes%20due%20to%20learning.

69    "Academic Education Can Positively Affect Aging of the Brain." Neuroscience News, January 2, 2022. https://neurosciencenews.com/education-aging-brain-19858/#:~:text=The%20benefits%20of%20good%20education,related%20cognitive%20and%20neural%20limitations.

70    Murphy, Emma. "What Are the Long Term Benefits of Learning?" Association of Learning, December 8, 2022. https://associationoflearning.com/what-are-the-long-term-benefits-of-learning/.

71    Harris, Brittany. "Auto Mechanic Achieves Dream of Becoming Doctor at 51 Pkg." Cleveland Clinic Newsroom, March 30, 2023. https://newsroom.clevelandclinic.org/2023/03/30/auto-mechanic-achieves-dream-of-becoming-doctor-at-51-pkg/#:~:text=CLEVELAND%20%E2%80%93%20Carl%20Allamby%2C%20MD%2C,Allamby.

72    Canfield, Jack. The Success Principles. New York, NY: Harper Collins, 2007. page 43

73    Weir, Malcolm, and Lynn Buzhardt. "Can Old Dogs Learn New Tricks?: VCA Animal Hospitals." VCA. Accessed February 1, 2024. https://vcahospitals.com/know-your-pet/can-old-dogs-learn-new-tricks#:~:text=Even%20though%20young%20pups%20may,for%20longer%20periods%20of%20time.

74    Ibid.

75    Berger, Susanna. "You Can Teach an Old Dog New Tricks, but Younger Dogs Learn Faster." Neuroscience News, February 3, 2016. https://neurosciencenews.com/learning-cognition-dogs-3566/#:~:text=Faster%20%2D%20Neuroscience%20News-,You%20Can%20Teach%20an%20Old%20Dog%20New,But%20Younger%20Dogs%20Learn%20Faster&text=Aging%20affects%20the%20cognitive%20abilities,at%20the%20Vetmeduni%20Vienna%20shows.

76    "Chaser, the Smartest Dog in the World." Chaser the border collie. Accessed February 1, 2024. https://www.chaserthebc.com/.

77    "Super Dog Knows 340 Words." Dogs Monthly, January 19, 2016. https://dogsmonthly.co.uk/2009/12/15/super-dog-knows-340-words/.

78    Betsy (Dog). https://en.wikipedia.org/wiki/Betsy_(dog)#:~:text=Betsy%20has%20a%20vocabulary%20of,and%20regards%20it%20as%20such.

79    Schrier, Charlie. "The Benefits of Learning by Doing: STRIVR Blog." Strivr, June 16, 2023. https://www.strivr.com/blog/learn-by-doing/#:~:text=When%20a%20hands%2Don%20approach,encouraged%20to%20learn%20by%20doing.

80    Cpd. "Importance of Repetition in Learning." The CPD Certification Service, September 15, 2022. https://cpduk.co.uk/news/importance-of-repetition-in-learning#:~:text=When%20stimuli%20are%20learned%20by,the%20performance%20of%20the%20skill.

81    Bernard, Sara. "Neuroplasticity: Learning Physically Changes the Brain." Edutopia, December 1, 2010. https://www.edutopia.org/neuroscience-brain-based-learning-neuroplasticity. =

82    Schatz, Itimar. Effectiviology. Accessed February 1, 2024. https://effectiviology.com/protege-effect-learn-by-teaching/#:~:text=Beyond%20improving%20your%20ability%20to,confidence%2C%20and%20improved%20leadership%20ability.

83    Seneca. "Seneca - While We Teach, We Learn." BrainQuote.com. Accessed March 12, 2024. https://www.brainyquote.com/quotes/seneca_405315.

84    Simplicio, Gea. "Teach to Learn Approach - Methodology Handbooks." WeSchool, May 25, 2023. https://www.weschool.com/methodology-handbooks/teach-to-learn/#:~:text=Teach%20to%20Learn%20is%20a,to%20effectively%20enhance%20learning%20outcomes.

**CHAPTER 13**

85    Dogged determination definition in American English | Collins English ... Accessed February 1, 2024. https://www.collinsdictionary.com/us/dictionary/english/dogged-determination.

86    Lacoste, Kristine. "5 Things to Know about Bulldogs." Petful, April 25, 2020. https://www.petful.com/breeds/breed-profile-bulldog/.

87    Aimon. "The Power of Bulldog Determination." Bullifieds Blog, October 15, 2022. https://bullifieds.com/the-power-of-bulldog-determination/.

88    Foundry, Zenrez. "Conjuring up Your English Bulldog Determination." The Foundry, May 4, 2012. https://thefoundryyoga.com/blog/2012/05/04/conjuring-up-your-english-bulldog-determination.

89    Ibid.

90    "Winston S. Churchill - Biography, Death & Speeches." History.com. Accessed February 1, 2024. https://www.history.com/topics/european-history/winston-churchill#section_6.6

91    "Sir Winston Churchill." History of Sir Winston Churchill - GOV.UK. Accessed February 1, 2024. https://www.gov.uk/government/history/past-prime-ministers/winston-churchill#:~:text=Winston%20Churchill%20was%20an%20inspirational,and%20from%201951%20to%201955..

92    Daniels, Anthony M, and J Allister Vale. "Did Sir Winston Churchill Suffer from the 'Black Dog'?" Journal of the Royal Society of Medicine, November 2018. https://www.ncbi.nlm.nih.gov/pmc/articles/PMC6243428/.

93    "Winston Churchill The Writer," America's National Churchill, 2024. https://www.nationalchurchillmuseum.org/winston-churchill-the-writer.html#:~:text=Winston%20Churchill%2C%20a%20gifted%20writer,supporting%20himself%20and%20his%20family

94    Cynewulf Robbins, Ron. "The Artist Winston Churchill," America's National Churchill Museum, 2024. https://www.nationalchurchillmuseum.org/the-artist-winston-churchill.html#:~:text=Over%20a%20period%20of%20forty,became%20half%20passion%2C%20half%20philosophy.

95    Winston Churchill Quote, Goodreads, Goodreads, Inc. 2024. https://www.goodreads.com/quotes/25265-continuous-effort---not-strength-or-intelligence---is-the

96    Fortgang, Laura Berman. "Got Dogg? (Dogged Determination That Is)." Now What?® Coaching, February 17, 2014. https://nowwhatcoaching.com/2014/02/17/got-dogg-dogged-determination-that-is/.

97    "Decide - Define Your Purpose and Decide to Cut off Everything Else." DECIDE - Define your purpose and decide to cut off everything else. Accessed February 1, 2024. https://www.meaningfulhq.com/decide-to-cut-off.html.

98    Fortgang, Laura Berman. "Got Dogg? (Dogged Determination That Is)." Now What?* Coaching, February 17, 2014. https://nowwhatcoaching.com/2014/02/17/got-dogg-dogged-determination-that-is/.

99    Hauglann, Maria Wulff. "The Amazing and True Story of Hachiko the Dog." Nerd Nomads, May 3, 2021. https://nerdnomads.com/hachiko_the_dog.

100    Margaritoff, Marco. "Why the Real-Life Story behind 'Rudy' Is Even More Inspirational than the Movie Depicted." All That's Interesting, August 9, 2023. https://allthatsinteresting.com/rudy-ruettiger.

101    Connell, Sara. "'dogged Determination,' Is the Key to Writing a Bestseller, an Interview with Authors Sara Connell & Sherrilyn Kenyon." Thrive Global, March 19, 2020. https://community.thriveglobal.com/dogged-determination-is-the-key-to-writing-a-bestseller-an-interview-with-authors-sara-connell-sherrilyn-kenyon%ef%bb%bf/.

**CHAPTER 14**

102    Hood, Dr. Julia. "The Benefits and Importance of a Support System: Highland Springs Clinic." Highland Springs, July 26, 2022. https://highlandspringsclinic.org/the-benefits-and-importance-of-a-support-system/#:~:text=Some%20of%20the%20best%20benefits,when%20we%20are%20in%20need.

103    **Mayo Clinic.** "Feeling less lonely, isolated, or judged." Mayo Clinic Stress Support Groups.

https://www.mayoclinic.org/healthy-lifestyle/stress-management/in-depth/support-groups/art-20044655#:~:text=Feeling%20less%20lonely%2C%20isolated%20or,skills%20to%20cope%20with%20challenges

104    Ibid.

105    **PubMed.** "Stress recovery with social support: A dyadic stress and support task." PubMed Article. https://www.ncbi.nlm.nih.gov/pmc/articles/PMC5633215/

106    Ibid.

107    Mech DL. Alpha status, dominance, and division of labor in wolf packs. Can J Zool. 1999;77(8):1196–203. 10.1139/z99-099.

108    **Highland Springs Clinic.** "The Benefits and Importance of a Support System." Highland Springs Clinic. Importance of a Support System | Highland Springs Clinic

109    "The Importance of a Strong Support System." Ultimate Medical Academy. Accessed February 1, 2024. https://www.ultimatemedical.edu/blog/importance-strong-support-system/.

110    "Manage Stress: Strengthen Your Support Network." American Psychological Association, 2022. https://www.apa.org/topics/stress/manage-social-support.

111    "Mentoring vs Coaching: The Key Differences and Benefits." PushFar, 2021. https://www.pushfar.com/article/mentoring-vs-coaching-the-key-differences-and-benefits/#:~:text=The%20Definitions%20of%20Coaching%20and%20Mentoring&text=A%20mentor%20is%20someone%20who,another%20to%20develop%20and%20grow.&text=A%20coach%20is%20someone%20who,them%20reach%20their%20full%20potential.

112    "Why You Should Consider an Accountability Partner - USCI." U.S. Career Institute. Accessed February 1, 2024. https://www.uscareerinstitute.edu/blog/why-you-should-consider-an-accountability-partner#:~:text=The%20purpose%20of%20accountability%20partners,make%20accountability%20relationships%20successful%2C%20though.

113    Finkelstein, Darren. "Why Should You Have an Accountability Partner?" LinkedIn, June 29, 2020. https://www.linkedin.com/pulse/why-should-you-have-accountability-partner-darren-finkelstein/.

114    Burris, Lela. "What's an Accountability Partner and Do You Need One?" Organized-ish, July 7, 2021. https://www.lelaburris.com/accountability-partner/#:~:text=Accountability%20partners%20come%20in%20two,professional%20motivation%20along%20the%20way.

115    Greenstreet, Karyn. "Napoleon Hill and Mastermind Groups." The Success Alliance, May 11, 2023. https://www.thesuccessalliance.com/blog/napoleon-hill-and-mastermind-groups/.

116    Lavinsky, Dave. "The Incredible Power of Mastermind Groups." Growthink, May 23, 2023. https://www.growthink.com/content/incredible-power-mastermind-groups#:~:text=In%20England%2C%20a%20mastermind%20group,The%20Lord%20of%20the%20Rings.

117    Ibid.

118    "Benefits of Participating in Mastermind Groups." OAT. Accessed February 1, 2024. https://www.oatext.com/benefits-of-participating-in-mastermind-groups.php.

119    https://www.thesuccessalliance.com/blog/mastermind-for-90-days/#:~:text=Most%20mastermind%20groups%20last%20at,six%20month%20or%20a%20year

120    Rolly. "How to Setup Your Mastermind Group." Eventual Millionaire, November 4, 2020. https://eventualmillionaire.com/how-to-setup-your-mastermind-group/.

## CHAPTER 15

121   Bundhun, Manish. "From Alpha to Omega." LinkedIn, 2021. https://www.linkedin.com/pulse/from-alpha-omega-manish-bundhun-/.

122   Ibid.

123   Mech DL. Alpha status, dominance, and division of labor in wolf packs. Can J Zool. 1999;77(8):1196–203. 10.1139/z99-099.

124   "Dogs Are Pack Animals." Animal Care Hospital of Reynoldsburg. Accessed February 1, 2024. https://www.ancarereyns.com/news-and-info/pet-care/dogs-are-pack-animals/#:~:text=Because%20dogs%20were%20domesticated%20from,leadership%2C%20structure%2C%20and%20protection; https://vcacanada.com/know-your-pet/dog-behavior-and-training-dominance-alpha-and-pack-leadership-what-does-it-really-mean.

125   Gallup, Inc. "Leadership Effectiveness: How to Be a Better Leader." Gallup.com, January 29, 2024. https://www.gallup.com/cliftonstrengths/en/356072/how-to-be-better-leader.aspx.

126   Maxwell, John C. 21 irrefutable laws of leadership: Follow them and people will follow you. New York, NY: HarperCollins Leadership, 2022.

## CHAPTER 16

127   Westlund, Karolina. "4 Compelling Reasons to Feed Treats at the Vet's." ILLIS ABC, November 20, 1970. https://illis.se/en/4-compelling-reasons-to-feed-treats-at-the-vets/.

128   Ibid.

129   Yin, Sophia A. Low stress handling, restraint and behavior modification of Dogs & Cats: Techniques for developing patients who love their visits. Davis, CA: CattleDog Pub., 2009.

130   Amy Morin, LCSW. "Improve Your Child's Behavior Problems with Positive Reinforcement." Verywell Family, September 2, 2022. https://www.verywellfamily.com/positive-reinforcement-child-behavior-1094889.

131   "Recognition Satisfaction: Globoforce Press Release." Workhuman Press, May 23, 2023. https://press.workhuman.com/press-releases/globoforce-survey-more-americans-satisfied-with-recognition-at-work/.

132   Carnegie, Dale. How to enjoy your life and your job: Selections from "how to win friends and influence people" and "how to stop worrying and start living." New York, NY: Pocket Books, 1974. Pgs. 89-96

133   "Reward System." The Reward Foundation, September 24, 2022. https://

rewardfoundation.org/brain-basics/reward-system/#:~:text=A%20reward%20is%20
a%20stimulus,than%20pain%20for%20motivating%20behaviour.

134    "The Pleasure Trap: Douglas Lisle at TEDx Fremont." YouTube, December 5,
2012. https://www.youtube.com/watch?v=jX2btaDOBK8&embeds_referring_
euri=https%3A%2F%2Fvideo.search.yahoo.com%2Fsearch%2Fvideo%3B_ylt%3DAwrO_
W4VKrxlWVMC05BXNyoA%3B_ylu%3DY29sbwNncTEEcG9zAzEEdnRpZAMEc2V-
jA3Nj%3Ftype%3DE210US7&embeds_referring_origin=https%3A%2F%2Fvideo.search.
yahoo.com&source_ve_path=Mjg2NjY&feature=emb_logo.

135    Perry, Elizabeth. "You've Earned It: Learn about the Benefits of Rewarding Yourself." BetterUp, 2022. https://www.betterup.com/blog/reward-yourself#:~:text=Biolog-
ically%2C%20rewards%20increase%20dopamine%20levels,reinforces%20specific%20
behaviors%20as%20worthwhile.

136    Walker, Susanna. "The Science of Hugging, and Why We're Missing It So Much during
the Pandemic | Susannah Walker." The Guardian, April 14, 2021. https://www.theguardian.
com/commentisfree/2021/apr/14/science-hugging-missing-pandemic-human-touch-skin.

137    Ibid.

## CHAPTER 17

138    Robinson, Lawrence. "The Benefits of Play for Adults." HelpGuide.org, June 21,
2023. https://www.helpguide.org/articles/mental-health/benefits-of-play-for-adults.htm.

139    Bekoff, Mark. "Playful Fun in Dogs." Current Biology, January 5, 2015. https://www.
sciencedirect.com/science/article/pii/S0960982214011221.

140    Bekoff, Marc. "When Dogs Play, They Follow the Golden Rules of Fairness." Psychology Today, 2019. https://www.psychologytoday.com/ca/blog/animal-emotions/201911/
when-dogs-play-they-follow-the-golden-rules-fairness.

141    "My Life in Full." Instaread. Accessed February 1, 2024. https://instaread.co/insights/
science-nature/how-dogs-love-us-book/xnuhk3017j#:~:text=In%20this%20way%2C%20
the%20Dog,experience%20for%20their%20human%20companions.

142    Ibid.

143    Kuschmider, Rebekah. "Cat Body Language: How to Understand Your Pet." WebMD,
2021. https://www.webmd.com/pets/cats/features/cat-body-language.

144    Ibid.

145    Ibid.

146   "View of the Well-Being of Playful Adults: Adult Playfulness, Subjective Well-Being, Physical Well-Being, and the Pursuit of Enjoyable Activities: The European Journal of Humour Research." View of The well-being of playful adults: Adult playfulness, subjective well-being, physical well-being, and the pursuit of enjoyable activities | The European Journal of Humour Research. Accessed February 1, 2024. https://europeanjournalofhumour.org/index.php/ejhr/article/view/Rene%20Proyer/Rene%20Proyer.

147   "The Best Medicine: Study Finds Laughter Is Good for Heart Health." The Guardian, August 27, 2023. https://www.theguardian.com/society/2023/aug/27/the-best-medicine-study-finds-laughter-is-good-for-heart-health.

CHAPTER 18

148   Campbell, Polly. "Why You Should Celebrate Everything." Psychology Today, 2016. https://www.psychologytoday.com/ca/blog/imperfect-spirituality/201512/why-you-should-celebrate-everything#:~:text=According%20to%20social%20psychology%20researcher,challenges%20that%20cause%20major%20stress.

149   "Why Celebrating Our Success Is Important." Sheffield Mind. Accessed February 1, 2024. https://www.sheffieldmind.co.uk/why-celebrating-our-success-is-important#:~:text=Celebrating%20our%20success%20is%20important%20because%20it%20increases%20motivation%20towards,do%20what%20they%20are%20doing.

150   Cheyette, Benjamin. "Why It's Important to Celebrate Small Successes." Psychology Today, 2021. https://www.psychologytoday.com/ca/blog/1-2-3-adhd/202111/why-its-important-celebrate-small-successes.

151   "6 Reasons Why You Should Celebrate Success." Brilliant Living HQ, March 18, 2018. https://www.brilliantlivinghq.com/6-reasons-why-you-should-celebrate-success/.

152   Asset, Creative. "Healthy and Happy: The Importance of Celebrating Life Milestones." Altura Learning, July 5, 2021. https://www.alturalearning.com/healthy-and-happy-the-importance-of-celebrating-life-milestones/#:~:text=Celebrating%20is%20good%20for%20people,pressure%20and%20relaxes%20tight%20muscles.

153   "Why Do Cats Sleep so Much?" Encyclopædia Britannica. Accessed February 1, 2024. https://www.britannica.com/story/why-do-cats-sleep-so-much#:~:text=The%20truth%20is%20that%20cats,especially%20older%20cats%20and%20kittens.

154   Cydcor. "The Importance of Rest and Relaxation." Cydcor, October 31, 2019. https://www.cydcor.com/the-importance-of-rest-and-relaxation/.

155   "The Importance of Rest and Relaxation for Successful Leadership." Touch Stone Publishers LTD, June 15, 2022. https://touchstonepublishers.com/worklifebalance/.

156   Staff. "Find Your Movement, Change Your Mood: The Emotional Benefits of Exercise." Ask The Scientists, July 4, 2022. https://askthescientists.com/exercise-mood/.

157   "10 Concentration Exercises to Improve Your Focus." BetterUp. Accessed February 1, 2024. https://www.betterup.com/blog/concentration-exercises.

158   "Why You Should Make Time for Self-Reflection (Even If You Hate Doing It)." Harvard Business Review, June 25, 2017. https://hbr.org/2017/03/why-you-should-make-time-for-self-reflection-even-if-you-hate-doing-it#:~:text=Reflection%20gives%20the%20brain%20an,inform%20future%20mindsets%20and%20actions.

159   "A Quote by Peter F. Drucker." Goodreads. Accessed February 1, 2024. https://www.goodreads.com/quotes/7346041-follow-effective-action-with-quiet-reflection-from-the-quiet-reflection.

160   2011-2024, (c) Copyright skillsyouneed.com. "Journaling for Personal Development: Creating a Learning Journal." SkillsYouNeed. Accessed February 1, 2024. https://www.skillsyouneed.com/ps/personal-development-journal.html.

# Permissions

Taken from *The 21 Irrefutable Laws of Leadership* by John C. Maxwell Copyright © 1998 by Maxwell Motivation Inc. Used by permission of HarperCollins Christian Publishing. www.harpercollinschristian.com

Used with permission from McColl, Peggy—Home—Peggy McColl

Used with permission from Taylor, Adam - https://www.linkedin.com/pulse/work-life-equation-understanding-your-talents-gifts-ksas-taylor-ccp/

Used with permission from Bekoff, Marc—*Dogs Demystified: An A-to-Z Guide to All things Canine* by Bekoff, Marc (amazon.com)

# About the Author

Dr. Teresa Woolard is a veterinarian, author, and personal development speaker who believes animals can teach us a lot about being better humans. Having studied and practiced veterinary medicine for twenty-two years, Dr. Woolard has written and sold 400,000 copies of her pet care books and consulted for the Canadian veterinary industry.

In addition, she has also built two award-winning brands in the hospitality industry, earning the Woman Entrepreneur of the Year award in her local community. As a leader, speaker, entrepreneur and consultant, Teresa can often be found giving back through teaching, inspiring, leading, and hosting numerous business, women, pet, and charitable events.

Armed with the perspective and understanding of animal behavior, as well as having a comprehensive scientific and sociologic understanding through two additional University degrees in Biochemistry and Gerontology, Dr. Woolard sees many analogies and similarities between animals and humans and how we can learn simple and fun approaches from our animal friends to achieve our goals in life, grow personally, and better our lives overall.

Dr. Woolard also enjoys kayaking, canoeing, running, and weight-training. She competed nationally in the Grand Masters Level Bikini Fitness competition in mid-life. An avid reader, Dr. Woolard lives with her husband in picturesque Northern Ontario, surrounded by nature.

For information on upcoming events, additional resources, and products that celebrate the human-animal bond, or to delve deeper into finding and using your gift with her course, *Dig. Leap. Play.* visit www.drteresawoolard.com.

# BOOK DR WOOLARD TO SPEAK AT YOUR NEXT EVENT!

CALL (705) 302-0327 OR BOOK HERE:

# Notes

Manufactured by Amazon.ca
Bolton, ON

39439413R00142